Sustainability in Business

David Hobson Myers

Sustainability in Business

A Financial Economics Analysis

David Hobson Myers
D'Amore-McKim School of Business
Northeastern University
Boston, MA, USA

ISBN 978-3-319-96603-8   ISBN 978-3-319-96604-5 (eBook)
https://doi.org/10.1007/978-3-319-96604-5

© The Editor(s) (if applicable) and The Author(s), under exclusive license to Springer Nature Switzerland AG 2020
This work is subject to copyright. All rights are solely and exclusively licensed by the Publisher, whether the whole or part of the material is concerned, specifically the rights of translation, reprinting, reuse of illustrations, recitation, broadcasting, reproduction on microfilms or in any other physical way, and transmission or information storage and retrieval, electronic adaptation, computer software, or by similar or dissimilar methodology now known or hereafter developed.
The use of general descriptive names, registered names, trademarks, service marks, etc. in this publication does not imply, even in the absence of a specific statement, that such names are exempt from the relevant protective laws and regulations and therefore free for general use.
The publisher, the authors and the editors are safe to assume that the advice and information in this book are believed to be true and accurate at the date of publication. Neither the publisher nor the authors or the editors give a warranty, expressed or implied, with respect to the material contained herein or for any errors or omissions that may have been made. The publisher remains neutral with regard to jurisdictional claims in published maps and institutional affiliations.

Cover pattern: © Melisa Hasan

This Palgrave Macmillan imprint is published by the registered company Springer Nature Switzerland AG
The registered company address is: Gewerbestrasse 11, 6330 Cham, Switzerland

Preface

Since the initial draft of this book, the world has been changed by the COVID-19 crisis. The crisis, however, serves to highlight the choices societies, businesses, and individuals make. The importance of a clear, concise, consistent, and convincing decision model is thus that more critical. The original aim of the book was to provide such a context for understanding the choices businesses make among their current and future shareholders, customers, employees, suppliers, and communities. The decision model structure presented here may help students and researchers make sense of the world and the choices they now face.

Sustainability is about ethics, philosophy, and social interactions. It is about how people approach each other and how they approach future generations. Economics is about choices and scarce resources. Business necessarily lies across both sustainability and economics. The choices highlighted in today's world of pandemic, geopolitical unrest, and economic uncertainty make that a very stark reality. The "green" movement has aimed some of its efforts towards reducing plastics and greenhouse gases. The pandemic has switched society from moving towards more reusable cups, grocery bags, and metal straws back to throwaway utensils to prevent the spread of the virus. On the flip side of the sustainability coin and the pandemic, we have seen a dramatic decrease in greenhouse gases due to the slowdown in economic activity. The advantages and disadvantages reflected in the response to the pandemic magnify the choices among current and future generations. The aim here is to provide a decision model for those in a business environment to make better informed choices.

Business and societal choices are about sustainability and the tradeoffs of current generations versus future generations. The model presented is based on the economics and finance of discounted cash flows with the addition of social distance as a discount factor. Social distance may mirror issues of empathy towards other groups. These measures provide a structure to understand those tradeoffs. Discount factors based on time and social distance remind consumers that a dollar tomorrow is less than a dollar today or a dollar to someone else is less than a dollar to us. How much less is based on the size of the discount factors.

Businesses and organizations have competing interests of their shareholders and other stakeholders. The economic foundations of a model based on social distance and time discounting extend the traditional discounted cash flow models of net present value to one of sustainable net present value. Such a structure for decision making provides a framework for thought and reflection about the business choices today and impacts those choices have on current and future stakeholders and on profits.

The objective of this book is to provide a text that more firmly roots the concepts learned in financial economics to the field of sustainability. Most textbooks in the arena of sustainability for businesses are from the management perspective and those in the economics literature most often focus more on macroeconomic goals such as societal goals. The aim here is to frame a discussion of sustainability for businesses or organizations and describe how to incorporate financial economic techniques in their decision-making process.

In the broadest sense, economics is about the allocation of scarce resources. The allocation of such resources is a question of choices. Choices of the consumption and allocation of scarce resources are being made by individuals, businesses, governments, and societies. Those choices are also central to sustainability. Choices have an impact for others whether the others are current generations or future. Those choices impact natural resources, shared resources, public, and private resources and the allocation of those resources. The framework adopted here provides a lesson in the concepts of utility maximization for sustainable businesses. Such lessons will provide researchers with a theoretical basis to approach research questions on sustainability.

Since choices impact others, society and government create norms and laws to protect members of society from each other. A responsible or sustainable business may simply be one that rises above the minimum societal standard. The decision-making process for any individual or organization becomes the understanding, rationale, and commitment to rise above those minimum standards. The simple goal to maximize shareholder wealth and meet all legal standards is a starting point for all businesses; sustainability is about going beyond. In particular, the next step is to understand the impacts on stakeholders present and future and the ethical and moral responsibilities of the organization. The approach necessitates being cross-disciplinary. An approach that will touch on philosophy, psychology, sociology, and political science in combination with the core financial economist approaches.

The format for each chapter is begins with an introduction to theoretical financial economic approaches that will be useful to creating more sustainable businesses. Examples of applying these approaches will provide further understanding of their advantages and disadvantages and the challenges of sustainable decision making. Each chapter ends with some discussion of societal and governmental roles within which organizations must operate.

The book is outlined as follows:

The introductory chapter lays out the case for the approach to sustainability in a financial economics framework. Discussion of different definitions of sustainability frames how those definitions affect the approach researchers, students, and business people take in determining their decision making. The contrast between traditional finance, Friedman's "Maximize Shareholder Wealth", and sustainability's double or triple bottom lines (environmental, social, and governance or profit, people, and planet) are laid out as the foundation for those discussions.

Chapter 2 introduces utility functions and concentrate on consumption for both the individual and society. Given the contrast between maximizing shareholder wealth and triple bottom lines, this chapter discusses utility functions with an emphasis on consumption models. The case for sustainability relies on social distance measures for the intergenerational and intragenerational transfers where the utility is a function of time value of money and social distance (Becker 1968 and Anderson and Myers 2017 and 2020). The discussion of utility is supplemented with discussions of Von Neumann Morgenstern Utility, transitivity, and one period models versus multi-period models as a means of discussing the framing of sustainability beyond traditional finance models.

Chapter 3 introduces a discounted cash flow approach to valuation of business ventures and highlights the key element of growth. The success of a business, sustainable or not, is dependent on its growth. Expansion of a business relies on expanding its markets' reach (demographics), its innovation or creation of new products, and its capital structure (leverage) within a legal and governmental framework. The interplay of social distance and demographics highlights the markets that a business may expand into as it builds customer relationships. Those relationships are based on the marketing of innovative sustainable products. The role of leverage on capital structure is described within a risk and return framework.

Chapter 4 introduces the reader to different ratings created and employed to measure sustainability. In building relationships with consumers for sustainable products or business, the trust and certification of sustainability is dependent upon measurement. The chapter discusses methods of certification through organizations such as LEED, Bloomberg's ESG, MSCI KLD, and other assurances of sustainability. The strengths and weaknesses of the measurability will be highlighted by the presence of false advertising known as "greenwashing."

Chapter 5 highlights the different legal structures for businesses and how sustainability fits with those structures. Corporate or business implementation strategies are covered. Implementation of structure and process in a sustainable framework highlights investments in products through a Sustainable NPV (SNPV) analysis or business structure. Differences in structure and form such as B-Corps, non-profits, institutions, partnerships, and corporations are discussed within the area of sustainability.

The final chapter is a review of the investment opportunities and strategies for sustainably minded investors. A review of the historical sustainable movement from ethical screens to socially responsible investment to mission related and impact invest investing is made to highlight changes in the costs of monitoring and investor interest which have created a greater interest in sustainable investing. Strategies from positive and negative screening to shareholder activism as well as best in class are discussed.

Boston, USA David Hobson Myers

Contents

1 **Introduction**   1
   Sustainability and Sustainable Goals   3
     Basic Economic Concepts and Their Awesome Power   4
   Model for Sustainability with the Concept of Social Distance   6
   United Nations Sustainable Development Goals (UN SDGs)   8
   Intergenerational and Intragenerational Transforms   10
   A Word of Caution on Economic Models   12
     Plato's Allegory of the Cave (A Strange Interpretation)   14
   Government's Role in Defining Limits and Generational Transfers   16
   References   17

2 **Economic Models**   19
   Transitivity and Binary Choices   21
   Stated Versus Revealed Preference   23
   Portfolio Choice   25
   Consumption Utility Model   25
     Consumption Utility Model Through Time   28
   Fisher Approximation (1930)   29
   Sustainability Utility Model   29
   Line in the Sand or a Spectrum of Options   31
   Binomial Pricing Models for Uncertainty   32
   Edgeworth Box and Societal Agreement   34

	Infinity and Finance	37
	Infinite Costs	37
	Infinity and Going Concern	38
	References	39
3	**Growth and Business Sustainability**	**41**
	Leverage	43
	Innovation	45
	Demographics	46
	Role of Government	48
4	**Certification of Sustainability**	**51**
	Trust and Acceptance in the Certification	52
	Measurability of the Certification	53
	Relationship of Certification to Social Distance	54
	Relationship of Certification to Stakeholders	54
	Sustainable Accounting Standards Board (SASB)	56
	Governmental Role in Certification	56
	Stability and Consistency	57
	Reference	59
5	**Corporate Implementation and Business Forms**	**61**
	Net Present Value and Discounted Cash Flows	62
	Payback Period	65
	Internal Rate of Return (IRR)	67
	Equivalent Annual Cost: Long Versus Short Term	68
	Sustainable Net Present Value (SNPV)	69
	Manipulations to the Cash Flow Model	72
	Business Forms	73
	Sole Proprietorships	75
	Partnerships	75
	Limited Liability Corporations: Private	76
	Limited Liability Corporations: Public	76
	Non-profits and Non-governmental Organizations (NGOs)	77
	B-Corporations	78
	Other Potential Advantages to Sustainable Corporations	78
	Governmental Roles and International Differences	79

	Refresher on General Finance Variables	80
	General Asset Pricing Models (APM)	80
	Capital Asset Pricing Model (CAPM)	80
	References	83
6	**Investment Implementation**	**85**
	Theory to Evolution in Practice	86
	History of Sustainable Screening	88
	Investor Types	91
	Individuals	92
	Pensions	93
	Endowments	93
	Foundations	94
	Governmental Role	95
	Choice of Sustainable Approach	96
	Constrained and Unconstrained Portfolios	97
	Climate Change Investing and Risk Management	98
	Portfolio Creation	99
	Other Investment Vehicles	100
	References	101
Epilogue		**103**
Index		**105**

List of Figures

Fig. 2.1	Mean-variance efficient frontier	26
Fig. 2.2	Budget constrained utility optimal two product indifference curve	27
Fig. 2.3	Binomial pricing	32
Fig. 2.4	Edgeworth box similar utilities	36
Fig. 2.5	Edgeworth box disparate utilities	37
Fig. 3.1	Impact of leverage on returns	44
Fig. 3.2	Leverage, capital structure, and bankruptcy	45
Fig. 6.1	Returns and social distance to non-shareholders (stakeholders)	95

LIST OF TABLES

Table 1.1	UN's millennium development goals (2000–2015)	8
Table 1.2	UN's sustainable development goals (2015–2030)	9
Table 1.3	UN SDGs, social distance, and discounting	13
Table 2.1	Intransitive preferences	22
Table 2.2	Regression coefficients of local versus organic attributes	24
Table 4.1	Rating criteria example	55
Table 4.2	Rating examples	56
Table 4.3	UN SDGs as a rating system for an organization	58
Table 5.1	Less sustainable option	67
Table 5.2	More sustainable option	67
Table 5.3	NPV versus SNPV decisions	70
Table 5.4	NPV versus SNPV decisions for charity	71
Table 5.5	Organizational structure with social distance	74
Table 5.6	Variables fixed and random	82
Table 6.1	UN PRI 6 principles	88
Table 6.2	Categories of qualitative screens from Anderson and Myers (2007)	91
Table 6.3	Investment approaches by implied social distances and information costs	92
Table 6.4	Investor type with social distance and returns	96
Table 6.5	S&P 500 top 10 versus 2 largest SRI mutual funds	100

CHAPTER 1

Introduction

Abstract The introductory chapter lays out the case for the approach to sustainability in a financial economics framework. Discussion of different definitions of sustainability frames how those definitions affect the approach researchers, students, and business people take in determining their decision making. The contrast between traditional finance, Friedman's "Maximize Shareholder Wealth," and sustainability's double or triple bottom lines (environmental, social, and governance or profit, people, and planet) are laid out as the foundation for those discussions.

Keywords Intragenerational transfers · Sustainability · Social distance · Business · UN SDGs

In preparing students as future business leaders and researchers for a sustainable future, it is important to lay out the arguments for and against sustainability that they will face. Those arguments center on definitions of sustainability and the battle lines that exist among the different stakeholders. This text is a step toward outlining those arguments from a financial economics standpoint. Most business sustainability texts are written from a management perspective. It is the goal here to frame the issues of sustainability against the backdrop of financial economics. Such a framework necessitates outlining the distinction between the classic finance approaches and the behavioral finance approaches to the issues and combining the result with sustainability.

In addition to the battle lines within finance, another hurdle that business leaders must be cognizant of is the wide range of definitions and therefore disagreements of what is "sustainability." Without agreement on what sustainability is, there is little possibility for agreement on solutions. Ultimately, it will be incumbent upon each reader to determine the best definition and solution for their work. It is the hope that this text will provide the tools to do so. This approach may be unsatisfactory to some. A. D. Roy (1952), the economist, best summarized this when he wrote "A man who seeks advice about his actions will not be grateful for the suggestion that he maximize utility." Yet this is exactly what this text intends to do. Utility functions, Chapter 2, are one of a financial economist's tools to explain human behavior. This is true whether one is a classical economist or behavioralist.

For the classical economist, Milton Friedman in 1970 set out the one and only directive for corporations (their utility function must include) maximizing shareholder wealth. If sustainability goals do not maximize shareholder wealth, then those goals are inappropriate for the firm. Since Friedman placed this stake in the theoretical ground of corporate behavior, there have been movements to justify changing that objective. Management scholars offer maximizing stakeholder wealth as an alternative to the traditional finance goal of maximizing shareholder wealth. If the stakeholders include the employees, the customers, and the community, then the theory is inching away from the shareholder toward stakeholders and a definition of more sustainable goals. Within stakeholders, future generations and current generations are included to be consistent with the goals of sustainability. In the summer of 2019, over 180 US CEOs announced that businesses had a responsibility to stakeholders not just shareholders. The stakeholders mentioned were employees, customers, suppliers, and the communities. This was reiterated on a broader scale at Davos in 2020.

For those looking for clarity and definitive answers, this text will not provide those. The approach taken here is that there are many approaches to sustainability. Individual solutions will be dependent on the organization/business and its stakeholders and the social distances to those stakeholders. The good news is that good students, researchers, and future leaders will recognize that solutions require thoughtfulness through critical thinking. The approach taken here should serve them well in all their endeavors.

The ability to create a more sustainable business or organization will rely on the ability to think beyond just profit and shareholders. It will require the ability to recognize how sustainability is dependent on the ability to view the impact of decisions on current and future generations. Those decisions reflect the economic choices that are made with respect to the consumption choices of current and future generations. Remembering that economics is the study of the allocation of scarce resources. The assumption is resources that are scarce today will be even scarcer in the future. In creating profitable businesses that are also sustainable will require changing human behavior and corporate behavior. Not an easy task to take on, but one that appears to be more pressing today as governments from local to national struggle with issues of climate change, pollution, and poverty. Remember that businesses, organizations, and individuals work within the societies that they belong and the rules and regulations within they reside. With that heady charge, we begin our journey.

Sustainability and Sustainable Goals

The best place to begin is to ground the discussion in a common definition of sustainability. There is a plethora of definitions for sustainability from financial sustainability to environmental sustainability. Economics centers on choice and choices have consequences to ourselves and others. As an example, the choices made by the ordering of the United Nations Sustainable Development Goals (UN SDGs: https://www.un.org/sustainabledevelopment/sustainable-development-goals/) presents the starting point for the first definition of sustainability. The UN SDGs will be enumerated in more depth later.

Other widely accepted definitions of sustainability include Landrum and Edwards' (2009),

> We will define sustainable business as one that operates in the interest of all current and future stakeholders in a manner that ensures the long-term health and survival of the business and its associated economic, social, and environmental systems.

And Sanders and Wood's Definition

"Positive social impact, a reduced negative environmental impact, and a positive economic impact." or "a business's contribution to social justice, environmental quality, and economic propriety is collectively referred to as the triple bottom line…or people, planet, profit."

All these definitions fit within the approach to sustainability of realizing that economic decisions affect current and future generations through the allocation of scarce resources. Thus, to be sustainable and to make those sustainable decisions in a manner consistent with the goals and objectives, the lessons of financial economics will be the guide.

Basic Economic Concepts and Their Awesome Power

The beginning of the twenty-first century has witnessed the awesome power of the simple economic concepts such as supply and demand with respect to oil prices, natural gas prices, and recycling. Additionally, the pandemic has changed decisions from future generations to current generations as societies have returned to plastic cups and bags to reduce further spreading of the COVID 19 virus. In both cases, to control or mitigate this power governments often jump into maintain sovereignty. The levers that they pull may be from central banks and interest rates to tariffs and trade controls. Even in their efforts to impact a more sustainable future, regulations and subsidies for electric vehicles, biofuels, and alternative power sources have subsidiary effects on both current and future generations. The role of societal norms expressed through governmental action decides the future of sustainability.

An example of the rippling effects of governmental intervention comes from the early years of the Obama administration. Efforts to increase the use of biofuels had downstream effects on food prices driven by both the increase in prices for corn and sugar from the substitution effects of increased demand for biofuels as well as protection effects from restrictions on support for farmers and agricultural policies.

The trade-offs between stakeholders or communities within current generations and across national borders have a profound effect on the international and global economic goals and progress between the developed and developing worlds. The UN SDGs highlight this tension both in terms of current versus future generations, but also within current generations between the developed and developing economies. Simple questions of fossil fuels and economic growth limits and targets have significant economic implications for all countries. One example is if to slow climate change, are developing economies sacrificing growth to future generations? These trade-offs among economies have been the key discussion points in most of the climate accords over the past few decades (Kyoto, Paris, ...).

Another example of this tension between developed and developing is China has become more stringent on the recycled materials they accept. In the past, they were willing to take recycled materials from the United States and Europe. The restrictions now in place in China and other countries have resulted in the cost of recycling to rise and are combined with a decreasing price of recycled materials as supply grows. Numerous accounts have surfaced of US cities abandoning recycling programs because the programs are no longer economically viable. The margins for recycling have become much worse.

Similar economic pressures are seen with oil prices and cars. Electric cars became more popular in the early part of this century as gas prices rose. With the economic decline after the Great Financial Crisis, gas prices fell and the demand for SUVs rose. Lower gas prices made SUVs more economical. More recently, the cost of electric vehicles has come down, and in combination with climate concerns, electric vehicles have become more popular, but an increase in electric vehicles pushes down demand for gasoline and thus prices. The dynamic will be the trade-offs (marginal costs and benefits) of climate concerns, electric vehicle prices versus lower gasoline prices, and affordability of internal combustion engines (ICE). There are also implications for funding of roads which are presently supported by gasoline taxes. To fund roads in an era of increasing electric vehicle world means transferring taxes from gasoline to electric vehicles.

The power of the marginal cost and marginal benefit rules of decision making is also evident in economics and sustainability. Returning to another of the supply and demand examples, it is the lowering of the marginal cost of production of natural gas that has increased its demand and decreased the demand for coal in electricity production.

Key economic concepts to remember from microeconomics:

1. Maximizing utility leads us to marginal cost equals marginal benefit and optimal decisions.
2. Time value of money and the marginal rate of intertemporal substitution assist in investment and consumption decisions.

3. Expected values are the probability of an outcome times the value of the outcome summed over all possible outcomes (the probabilities must sum to one).

Model for Sustainability with the Concept of Social Distance

After the first choice of defining sustainability with the UN SDGs, the second choice is to use the framework of "social distance" to discuss the issues of sustainability within the UN SDGs. Social distance was introduced by Becker (1968) to discuss the consumption choices within a family unit. Given the theory was in the 1960s, the framework was a father figure as the breadwinner making decisions for the family. The argument was that a dollar for the father was worth a dollar for family or any individual in the family. The conclusion was that in the family unit there was no or zero social distance. Anderson and Myers (2018) expand on this analogy toward more of a utilitarian interpretation of each consumer has known social distances and thus discount rates to others.

In approaching the decisions made by individuals, groups, and society for their own benefit, the model of maximizing wealth is greatly lacking in its ability to explain pro-social behavior. Economists have struggled with altruistic behavior. A model that is more widely applicable includes the addition of social distance. Social distance is a measure of how a decision maker relates to others. The impetus for this goes back to Becker and social distance within a family unit. For the family unit, there is no distance within the nuclear family. A dollar to one member is worth a dollar to all members. The concept of social distance becomes a discount factor as one moves outside the nuclear family. The further the social distance, the greater is the discount factor.

By employing social distance as a discount factor, Anderson and Myers leverage Becker to create a framework to corral the disparate definitions in the world of sustainability into a model of economic growth, consumption, and social distance. The model is employed throughout the text as the framework for sustainable decision making. The model will be discussed in more depth later both within the context of general consumption models as well as applications to sustainable decisions in organizations as an extension of net present value criterion. As a base case, social distance will be applied to the United Nation's Sustainable Development Goals as our broadest definition of sustainability. Social distance will be the metric to explain potentially the ordering of the UN SDGs.

Having begun my financial career with a Japanese brokerage firm in the 1980s and living in a company dorm, my first inkling of social distance as a way of explaining behavior came from that experience. A background for the social distance measure is from the structure and interactions within the Japanese society. If viewed as a series of concentric circles, Japanese have zero distance within the nuclear family as well, but distances may be seen to increase as one moves to family, school mates, college mates, business mates, to Japanese as a whole. The word for foreigners, "gaikokujin," is literally "outside country person." A poorer analogy is the solar system and the distances from the sun. Not only is Pluto's distance large, but its status as a planet has been debated. Even the current political debates of America First or populist movements in Europe may be seen as increasing social distance to those not in one's particular group.

The Economist writes about efficient charity that is also consistent with the model of social distance. Those making decisions about charity want to ensure who the money is going to help. Those who have the shortest social distance or smallest discount rate to the donor are the targets of the donations. Giving or having the greatest impact (utility) for the donor comes from getting the money to those with the shortest social distance. Efficiency of the donation or inefficiency may also be measured by monies going to those with greater social distance as being seen as more inefficient. That is if the money goes to not direct mission-related individuals versus mission-related individuals it is interpreted as being inefficient. Mission-related individuals as the target of the charity have lower social distance and thus greater utility to the donor.

Within the added dimension of social distance, it is much easier to discuss the range of definitions surrounding sustainability. The model of social distance also firmly plants the idea of sustainability within a social justice framework matching the SDGs of how decisions affect others and how much does it matter to the decision maker. Take, for example, the acronym ESG meaning Environmental, Social, and Governance. Environmental impact is about the effects on future generations and the current generation and the impact on that generation's health and well-being and their consumption goals. There will be more covered about consumption goals in Chapter 2's discussion of utility. Social goals are the impacts on other groups as well as governance's impact on the corporate culture and decision making. Environmentalists or social justice advocates can easily be categorized as having lower or shorter social distance to others, where those others are other groups (ethnic, racial ... or future generations). Short social distance translates to a lower discount rate which creates more value to the donor. A key to the social distance approach is to translate environmental issues as decreasing the social distance to future generations which creates more value to the decision maker. In modeling the utility functions in terms of social distance within a consumption-based utility, future generations' ability to consume at or near the same levels requires that the environment is sustained from one generation to the next. The best historical example of this is the Iroquois Nation belief in making decisions out to the seventh generation. One interpretation in light of social distance would be there is zero social distance to seven generations or they value all individuals out seven generations equal to the current generation.

United Nations Sustainable Development Goals (UN SDGs)

With the first two decisions having been made, we move now to meld the two together (1) sustainability defined with the UN SDGs and (2) social distance to consumption-based utility models for explaining and examining behavior. Having defined the UN SDGs as the definition of sustainability for this text, the United Nations Sustainable Development Goals (SDGs) will be examined through the lens of social distance.

The sustainable development goals were created as a follow on to the Millennium Goals which ran from 2000 to 2015 (see Table 1.1). The SDGs frame sustainability in a social justice and environmental agenda (see Table 1.2). The beneficiaries are global citizens of current and future generations. The aim is to lower the social distance for all global citizens

Table 1.1 UN's millennium development goals (2000–2015)	1. Eradicate extreme poverty and hunger 2. Achieve universal primary education 3. Promote gender equality and empower women 4. Reduce child mortality 5. Improve maternal health 6. Combat HIV/AIDS, malaria, and other diseases 7. Ensure environmental sustainability 8. Global partnership for development http://www.un.org/millenniumgoals/

Table 1.2 UN's sustainable development goals (2015–2030)

http://www.un.org/sustainabledevelopment/development-agenda/
1. No poverty
2. Zero hunger
3. Good health and well-being
4. Quality education
5. Gender equality
6. Clean water and sanitation
7. Affordable and clean energy
8. Decent work and economic growth
9. Industry, innovation, and infrastructure
10. Reduced inequalities
11. Sustainable cities and communities
12. Responsible consumption and production
13. Climate action
14. Life below water
15. Life on land
16. Peace, justice, and strong institutions
17. Partnerships for the goals

today and the future. The SDGs cross-national, ethnic, racial, socioeconomic, and generational boundaries. The first ten goals are aimed to the current generation and raising the economic level of the poor and disadvantaged. Goals 11–15 are more aimed toward environmental sustainability and thus future generations.

The UN SDGs represent an improvement on the Millennium Development Goals from 2000. Later chapters on certification and measurement highlight that with sustainability there will be continued improvement with better metrics and measurement of the metrics.

The simplest approach to understanding the UN SDGs in the context of social distance is to realize that all goals represent impact on certain populations and the need to make decisions in an efficient manner getting the assistance to groups with the lowest social distance or most immediate need. This is why the economic discussion centers on intergenerational and intragenerational transfers. By having an impact on certain populations, the goals are consistent with social distance measures and the time value of money to whom benefits and what the societal goals are for those groups, present and future. In this approach, strengthening others' ability to live healthy and productive lives is a result of reduced social distance and increased efficiency. Research has shown that productivity/economic growth is aided by better-educated and healthier workforces of both genders.

All the UN SDGs may be categorized and grouped by social distance and time across the relevant stakeholders. They are a balance between economic growth across the globe and the social distance to current and future generations. One view would be businesses will grow more if there are a larger number of consumers and wealthier consumers. To increase the demographic of wealthier consumers (witness Chinese economic growth from an increased middle class), it is incumbent on economic agents to be healthier and thus longer living consumers. To maintain that goal into the distant future, then providing a more sustainable future is necessary. The business, however, must be interested in the long, long term. Short-term profits and goals would be unsustainable. Now the business has recognized the importance of sustainable development goals.

The choice of the goals and the role of the United Nations imply a societal utility function. While this is an area of economics that has been long debated and will continue to be, it is not an area that will be debated here. The Anderson and Myers paper addresses more the individual or business utility function not societal. The interest is in how do individuals, businesses, and organizations make decisions to maximize their utility in the presence of their consumption and others through social distance measures. While future research and debate may continue to delve into social welfare functions, the biggest advantage future researchers will have is the use of big data and computing power. Big data and power to analyze it provides a straightforward approach to examining the aggregate decisions of the individuals of a society in a revealed preference approach.

Intergenerational and Intragenerational Transforms

The UN SDGs highlight that on a societal level there are questions of intergenerational and intragenerational challenges. The first few goals are more about intragenerational transfers, raising people out of poverty and squalor and ensuring basic human needs. The latter goals, especially those about the environment, are more about intergenerational transfers, ensuring that later generations may maintain a comparable level of happiness and consumption as today's generation. The economic literature about societal savings and consumption that has been leveraged for environmental economics begins with Ramsey (1928). Of course, the decision-making framework is one of discounting future consumption. This context of when the choice to consume resources provides the leap to financial economics and discounted cash flows. The key variable within all of these discussions is what is the appropriate discount rate and whether the rate is in isolation or combination with a social distance measure.

As the current generation tackles issues of climate change, societal inequalities, and economic growth, the key variable remains how one discounts the consumption of future generations. The Anderson and Myers (2018) approach leveraging Becker's (1968) social distance aids in the individual and business perspective in decision making by using a social distance discount rate to provide the framework. It will be much easier to answer the questions of sustainability for a business than it has been for a society given the shorter-term objectives of most firms. This also sheds a spotlight on the problems of societal decisions.

Even in C. S. Lewis (1947), there is a discussion of the power of present generations over nature and future generations by the decisions they make. Later in the environment economics literature, the language becomes starker as "dictatorship of the present versus the dictatorship of the future" (Daley and Townsend 1992). Given that overconsumption in the present may have a huge impact on the ability of future generations to consume or even survive, this generation is dictating to them what their lives will be like. Similarly, if this generation were to drastically reduce its consumption for the complete benefit of future generations then that would be a dictatorship of the future on the present generation. The environmental economics literature argues for very, very low societal discount rates to reflect the importance of sustainability on the billions living in the future (Stern [2007] and Arrow et al. [2013]). As businesses, organizations, and even as individuals, sustainability in business will be about finding that middle ground for the benefit of present and future generations. To interpret how the middle ground is found will be heavily dependent on the choice of discount rates and social distance. Rates in combination will be much higher than those societal rates advocated in environmental economics. This helps explain why businesses may not be the most appropriate place for societal goals and why the "Sustainable Net Present Value" with social distance is a more appropriate model for sustainable businesses.

Returning to the UN SDGs will provide a simple example of combining discount rates and social distance measures to explain or interpret the ordering of the goals. The order of the goals is assumed to be only a function of time discount and social distance. For the first 4 goals, the assumption of time is nearly immediate or 1 year. For goals 5–10, the assumption is of intermediate goals and time of 5 years. Finally, for the more distance goals, time is assumed to be 15 years since the UN SDGs are to be measured from 2015 to 2030. Overlaying the time assumption, the next assumption is a rank ordering for social distance within each of the time frames. The resulting implied discount factors and implied discount rates provide the same rank ordering as the UN SDGs. The key takeaway is that there is a combination of time and social distance to the value of a dollar today toward the goals. That is a $1 is worth $0.943 toward the no poverty goal and only about 24 cents to the life on land goal. Environmental economics argues in favor of a very low to zero discount rate to recognize the number of individuals in future generations to provide for sustainability. The aim here is not to generate social utility decisions for a society, but to provide a methodology for current individual organizational decisions to understand the implied social distance and the consequence of that implied social distance on the value to intergenerational and intragenerational transfers (Table 1.3).

The math behind the table is the implied discount factor is $1/(1 + R + \delta)^T$ with the time value of money held constant, R= 5%.

A Word of Caution on Economic Models

Having outlined the definition of sustainability via the UN SDGs and introduced social distance, the next step is to provide some background to the economic assumptions necessary for analysis of sustainability decisions. In the discussion of model choices, most of the introductory models are single-period models with certainty. That is a decision is made today and the outcome is measured one period out. For example, time value of money, discount rates, are based on a dollar today versus a dollar tomorrow, but tomorrow could be any period in the future. While models of certainty and single periods are a nice beginning, life and decisions are uncertain and decisions today will be impacted at many intermediate periods before the "tomorrow" is reached.

The future is uncertain and thus most planning by businesses, organizations, and individuals must be dealt with under uncertainty. Terms such

Table 1.3 UN SDGs, social distance, and discounting

UN SDGs	Time (T) in years	Social distance (δ)	Implied discount factor
No poverty	1	0.01	0.943
Zero hunger	1	0.02	0.935
Good health and well-being	1	0.03	0.926
Quality education	1	0.04	0.917
Gender equality	5	0.01	0.747
Clean water and sanitation	5	0.02	0.713
Affordable and clean energy	5	0.03	0.681
Decent work and economic growth	5	0.04	0.650
Industry, innovation, and infrastructure	5	0.05	0.621
Reduce inequalities	5	0.06	0.593
Sustainable cities and communities	15	0.01	0.417
Responsible consumption and production	15	0.02	0.362
Climate action	15	0.03	0.315
Life below water	15	0.04	0.275
Life on land	15	0.05	0.239

as risk and return and expectations are employed. These have a serious impact on the ability to make sustainable decisions as well. Consequently, good critical thinking models for decision making are even more necessary in framing the sustainable business decision. The tools an economist or statistician employs to handle expectations and uncertainty are statistical and probabilistic. Terms such as "mean" for expectations or "standard deviation" for risk are typical in the discussion of input variables in modeling.

Anyone familiar with financial modeling will be familiar with statistics and the role that many statistical assumptions play in those models. Nicholas Taleb in **Black Swans** attacks the simplifying assumption of normality in the Capital Asset Pricing Model (CAPM) and the Black–Scholes Option Pricing Models. It is easy to agree that the world of asset pricing is not simply normally distributed, the beautiful bell curve of symmetric probability distributions. In the same vein, there are and will be many shortcomings to the models and analyses presented here about sustainable decision making.

Incumbent on all decision makers is the recognition of the shortcomings of their models and their assumptions. The solution is critical thinking through a clear, concise, consistent, and convincing explanation of the power of the decision-making tools and assumptions. Many of the models presented in this text will be single-period models that are a decision today measured for an outcome in some tomorrow (time 0 to time 1). Many of the models assume that we know things with certainty when it is unlikely that we do. This is not too dissimilar from Secretary of Defense Rumsfeld's "there are things we know, things we don't know…" quote about weapons of mass destruction.

Plato's Allegory of the Cave (A Strange Interpretation)

Given people's desire for certainty and that in terms of the future, there is no certainty, it becomes incumbent upon decision makers to come to grips with the uncertainty. Plato's Allegory of the Cave is that man in the cave only saw shadows of real objects and not the objects themselves. The strange interpretation of the allegory is that our observations are measured with error (the shadows) and thus uncertainty. One way that financial economists deal with issues of uncertainty is reflected in the assumptions of statistical models. Some simple assumptions of an Ordinary Least Squares (OLS) regression are that the errors are on average zero, uncorrelated, and normally distributed. These are some of the same assumptions in the Capital Asset Pricing Model.

The discussion of uncertainty is not new and definitely much older than statistical models and the Capital Asset Pricing Model. One approach to adopt is an approach that probably represents a bastardization of Plato's Allegory of the Cave. The interpretation is that the truth or "true value" in finance is unobservable. This interpretation becomes very important in teaching about valuation since more often it is the process and assumptions that require most of the attention of the analyst and not a single value. The result is an understanding that value, as it is observed in market prices, is the true value plus some random error. Almost all the efficient markets tests can be categorized as tests also having expected errors equal to zero and that the errors are uncorrelated (tests of the random walk hypothesis).

The advantage of this "everything is measured with error" approach is that analysis of models and the resulting decisions are an analysis of reducing the variance of the errors and thus increasing the confidence in the decision. Mathematically or statistically, the everything is measure with error is written as:

$$\tilde{V}_{observed} = V_{truth} + \tilde{e}$$

With the statistical tests of the average error is zero and the errors are uncorrelated through time as:

$$E(\tilde{e}) = 0, E(e_t, e_{t+1}) = 0$$

One disadvantage of averages or expected values is that no one is average. A player who hits the dartboard on the top and bottom on average hits the bullseye. A true statement, but not a very helpful one. When a billionaire walks into a room of 9 average or non-billionaire people, the average person just rose at least $100 million dollars richer. This highlights the quote often attributed to Mark Twain, "Lies, Damn Lies, and Statistics." Statistics infuse almost all of finance and economics. Statistical tests of models rule the analysis and acceptance of research findings and are still applicable to models of any future, sustainable or not.

The efficient market hypotheses and statistical tests are tests of whether information is randomly generated; the errors on average would be zero. Sometimes the market prices the information too high and sometimes too low, but on average zero. Random walk extends the statistical test to whether the errors are correlated through time. If the market was up last period will it be up the next. Is there or is there not predictability in the market and can investors take advantage of it? Given how hard this is to do in the investment world, think now about how much harder it will be in terms of sustainable business decisions. Do not concentrate on the perfect answer, but spend time analyzing the information, the models, and the inputs to make better decisions, where better decisions statistically have more confidence through lower variability and better predictability.

Probably an even more important conclusion from the understanding of uncertainty and measurement error is the ability to address criticism of sustainable decisions as not having certainty. This is particularly true in terms of addressing and understanding the externalities of decisions on different stakeholders. How do carbon emissions and greenhouse gases affect current and future generations? How are those impacts quantified? In a discussion of the subsidies of the US government to the oil industry, the estimates range from roughly $30 billion to $500 billion. Whether you accept either number or how it is built into your decision making will also be a function of your social distance to current and future stakeholders. Think back to the discount factors implied in the UN SDGs (Table 1.1 and Table 1.3), if the ordering is correct then a change in social distance may be paired with a change in time. In the oil industry subsidies assumptions, the impact on one's decision may be driven by either a higher subsidy assumption or a lower social distance, both would result in the decision maker responding more to stakeholders impacted by carbon emissions.

Government's Role in Defining Limits and Generational Transfers

It is obvious that legal standards differ across governmental and societal lines and through time. From Prohibition in the earlier part of the twentieth century to the recent movement to legalize marijuana exhibits how laws and norms may change through time and across state and national boundaries. Environmental regulations change under different administrations. Safety standards may also change. The decision to follow or adapt to new standards may have a dramatic impact on a business's profitability and competitiveness. Unsurprisingly, businesses lobby for policies beneficial to them.

Following the global response to issues of climate change and the different governmental decisions exhibited in regulations, laws, and subsidies, highlight the societal decisions made by different social distance to current and future generations. Should current economic growth be sacrificed and thus the wealth and well-being of current generations for the good of future generations and the ability to be sustainable. The COVID-19 pandemic has given clear insight into the choices of economy, current, and future generations well-being. Quick decisions have been made to return to throw away coffee cups and plastic bags to reduce the virus spreading. It is incumbent on the decision maker whether they are individuals, businesses, organizations, or even governments to understand the impact the decision has on current and future generations.

Those decisions may be political or economic and the decisions will not be in isolation of the political/governmental or economic realm. Think about businesses efforts to change regulation or avoid regulation. The choice to actively lobby for changes in regulations is also an important business decision. The impact of the decisions may have dramatic effects across society.

Recent adoption on carbon emissions and electric vehicle decisions of different countries in Europe highlights different social distances and thus social discount rates toward future generations. The different target years give an indication of the differing social distance or importance of climate action. European targets by city or country range from 2025, 2030, and out to 2050.

REFERENCES

Anderson, A., & Myers, D. H. (2018). Sustainability: Discounting the Future, Social Distance, and Efficiency Effects, *Moral Cents*, (Winter/Spring).

Arrow, K. J. et al. (2013). How Should Benefits and Costs Be Discounted in an Intergenerational Context? Working Paper Series 5613, Department of Economics, University of Sussex Business School.

Becker, G. (1968, March/April). Crime and Punishment: An Economic Approach. *Journal of Political Economy*, 76(2), 169–217.

Daly, H. E., & Townsend, K. N. (Eds.). (1992). *Valuing the Earth, Economics, Ecology, Ethics*. Cambridge, MA: The MIT Press.

Landrum, N. E., & Edwards, S. (2009). Sustainable Business: An Executive's Primer, *Business Expert Press, LLC*.

Lewis, C. S. (1947). *The Abolition of Man*. New York, NY: The MacMillan Company.

Ramsey, F. P. (1928, December). A Mathematical Theory of Saving. *The Economic Journal*, 38(152), 543–559.

Roy, A. D. (1952, July). Safety First and the Holding of Assets. *Econometrica*, 20(3), 431–449.

Sanders, N. R., & Wood, J. D. (2014). *Foundations of Sustainable Business: Theory, Function, and Strategy*. Wiley: Hoboken, NJ.

Stern, N. (2007). *The Economics of Climate Change: The Stern Review*. U.K. Cabinet Office—HM Treasury.

Taleb, N. (2008). *The Black Swan: The Impact of the Highly Improbable*. Penguin Books Ltd. Penguin Random House: New York.

CHAPTER 2

Economic Models

Abstract This chapter introduces utility functions and concentrates on consumption for both the individual and society. Given the contrast between maximizing shareholder wealth and triple bottom lines, this chapter will discuss utility functions with an emphasis on consumption models. The case for sustainability will rely on social distance measures for the intergenerational and intragenerational transfers where the utility is a function of time value of money and social distance (Becker 1968 and Anderson and Myers 2018). The discussion of utility will be supplemented with discussions of von Neumann–Morgenstern Utility, transitivity, and one-period models versus multi-period models as a means of discussing the framing of sustainability beyond traditional finance models.

Keywords Utility functions · Sustainability · Stakeholder value · Social distance

In this chapter, economic models for decision making are presented in the context of sustainability. Consumption-based utility models (as opposed to production based) are the primary models presented. Central thesis to consumption models is that through time discounting future consumption is required. Future consumption is dependent on the current generations' consumption patterns. Obviously, for sustainability the impact of current consumption on future generations' consumption is paramount to sustainable decisions. Discounting future consumption is based on the trade-off of consuming today versus in the future. The discount rate is the intertemporal marginal rate of substitution. Sustainability is dependent on the trade-offs of consumption today versus consumption by future generations. Remember economists are interested in individuals maximizing their utility. The example from the previous chapter highlighted the United Nations' SDGs and their ordering as an examination of their utility function based on discounting through the time value of money and social distance factors.

Consumption-based models refer to choices today among consumption goods given their relative prices and present consumption versus future consumption. The choice of consumption today versus consumption tomorrow is critical to issues of sustainability. Not only because of an individual's choice of delayed consumption, but because reduced and better consumption by today's generation will have an impact on the ability of future generations to be able to consume. Once a better understanding of consumption-based utility models of the individual has been reached, the discussion moves to focus on sustainability with economic tools, modeling, and business in the context of stakeholders.

Besides consumption and utility, models of expectations and the associated probabilities are addressed in the context of binomial pricing. Given that the future is uncertain, one way to address that uncertainty is modeling of the probabilities of different outcomes. The simplest is the probability of good and bad outcomes, a binomial model. To expand the number of outcomes through time, a binomial model can be expanded with more steps or into a binomial tree (a model common to option pricing and decision sciences).

Given the contrast between maximizing shareholder wealth and triple bottom lines (People, Planet, and Profit), the case for sustainability relies on utility that is a function of time value of money and social distance. The discussion of utility is supplemented with discussions of von Neumann–Morgenstern Utility, transitivity, and one-period models versus multi-period models as a means of discussing the framing of sustainability beyond traditional finance models. Finally, consideration is given to choices of stated preferences versus revealed preferences and how to determine society's and individual's choices of sustainable decisions.

From its simplest form, a utility function attempts to model the choice behavior of a rational economic agent. Economics is about choices and in particular choices among scarce resources. Individuals, groups, and societies are having to make decisions across shared and scarce resources as well—air, water, and land. The quality of those shared resources affects the ability to enjoy them and affects the health and well-being of the members of that society. Environmental concerns are concerns of who, both current and future generations, is able to enjoy those resources and at what quality. If one has a shorter social distance to others, then the environmental concerns will be heightened in one's utility function. How effective sustainability efforts are is measured by transfers of costs and benefits to stakeholders. In summary, the following economic concepts, transitivity and binary choices, stated and revealed preferences, consumption utility models and infinite horizons and the firm are explored in the remainder of the chapter.

Transitivity and Binary Choices

The ability of a simple utility model to capture the complex nature of human decision making is limited. The reason to even attempt such an exercise is to provide a methodology for decision making that may lead to better decisions where better decisions mean better predictions through consistent assumptions. In the discussion that follows, the advantages and disadvantages of the models will be highlighted as a means to point the student and researcher in a direction to understand the strengths and weaknesses of the modeling and decision process.

In von Neumann–Morgenstern utility, one important assumption is that choices/preferences are transitive. If A is preferred to B and B is preferred to C, then A must be preferred to C. For a rational decision maker, this seems intuitive. In fact, it is easily shown that a decision maker without transitive preferences may be easily taken advantage of by someone else. A simple example of non-transitive preferences and the ability to take advantage (not ethical, but mathematical) of someone with such preferences is illustrated below.

> Assume an individual prefers alcohol to hamburgers (A > B), and hamburgers to cigarettes (B > C), but cigarettes to alcohol (C > A, where > implies preference). They are indifferent among their choices such that they would be willing to trade
>
> - alcohol for hamburgers and two dollars, A = B+$2
> - hamburgers for cigarettes and one dollar, B = C+$1;
> - and cigarettes for alcohol and $3, C = A+$3.

Table 2.1 Intransitive preferences

Time	Initial item	Trade	Cash
0	Cigarettes	Burger + $1	$12
1	Burger	Alcohol + $2	$11
2	Alcohol	Cigarettes + $3	$9
3	Cigarettes	Burger + $1	$6
4	Burger	Alcohol + $2	$5
5	Alcohol	Cigarettes + $3	$3
6	Cigarettes		$0

Let this consumer start with a carton of cigarettes and $12. First, they trade one burger for the carton of cigarettes and $1, followed by trading a bottle of alcohol for the burger and $2, next trade a carton of cigarettes for the alcohol and $3. This continues until they have no money left. People with irrational or intransitive preferences may be taken advantage and thus become economically irrelevant. This is not ethical, but mathematically possible. (Table 2.1)

The simple transitive preferences make it evident that transitive preferences are important for consistent and predictable choices. However, preferences may and do change over time for the economic agent. As circumstances change over time, a weakness of a single-period model is that it does not reflect how preferences may change. Thus, transitivity assumption may only hold in single-period models. Choices through time that may appear inconsistent, irrational, and non-transitive are simply reflecting the complexity of changing preferences through time. One way that economists and econometricians get around the issue is to define sufficiently long-time periods as a single period. Practically, this is driven as much by the theory as it is by the data. Historically, data on economies such as production and growth have been measured quarterly or annually. All the flows during a "single period" are assumed to have happened at once. Given advances in computing and thus data collection and the consequential growth in "Big Data" analytics, it is possible to dramatically shorten the length of a single period. Businesses with point of sales systems may look at daily or weekly data. Investors looking at Twitter, Google, or Facebook trends may look at Internet social media measures in even smaller increments (hourly or by the minute).

In the context of social distance, a leader of the country may always put their country first which translates as near zero social distance for their country and moving towards infinite social distance for other countries. If the choice is binary and single period then yes this is the correct interpretation. As the analysis moves to a more dynamic choice situation, this may not make sense. Situations with multiple choices or portfolio choices versus binary choices need different models of behavior. Political statements of "you are either with us or against us" or "American First" beliefs make good slogans, but are oversimplified like our single period transitive preferences. There is a similar dynamic with multi-period choices. One may prefer broccoli to Brussels sprouts, but after a week, month, or year of only broccoli, Brussels sprouts may not look so bad.

The dual choice issue, with us or against us, has been translated into for sustainability or against sustainability. Most sustainability issues are better framed as consumption goods with multiple aspects to them. This will ultimately lead us to more of a portfolio choice model in which the trade-offs are among a group of choices with utility defined in terms of consumption and social distance. To progress toward the goal of portfolio choice, we begin with the types of data available to analyze the choices that are being made today. Continuing in the preference and transitive framework, it will be important to differentiate between stated preferences and revealed preferences.

STATED VERSUS REVEALED PREFERENCE

For any business decision, it is helpful to understand who the stakeholders are. Stakeholders may be employees, customers, suppliers, shareholders, and the community beyond. Market research may play a large role in defining the prospects for a new product or market. A simple beginning may be to conduct surveys of potential stakeholders. These surveys may be representative of stated preferences and not true or revealed preferences. Stated preferences and survey biases are potentially a result of the interviewee projecting what they believe the interviewer is looking for in an answer. When students in a sustainability course are asked about their preferences, there may be a bias in favor of sustainability that would not be evident in asking just finance students, that is the unwelcome news. The better news is that with more and more data about consumers available through point of sales systems or data from Google and Facebook or other Internet sources on behavior of users, the future business person will have better (never perfect) data on revealed preferences. Does the individual who says they are sustainable make choices that are sustainable? Does the politician vote the way they said they would during the campaign? These are two simple examples of the potentially conflicting information from stated and revealed preferences. The 2019 Nobel Prize in Economics awarded to Abhijit Banerjee, Esther Duflo, and Michael Kremer highlights this with choice of randomized trials as an approach to get closer to unbiased and revealed preference analysis.

The advent of "big data's" impact on consumers' revealed preferences and predictability for a business is evident in a simple example using only four observations instead of the potentially millions of observations seen through Internet searches or point of sales data.

Even from a simple example, the power to tease out the value of consumer preferences is evident. Assume a grocery store which follows the purchasing patterns of certified organic apples and non-certified organic apples from local producers and non-local producers. The prices of the four different types of apples are given in the table below (Table 2.2). A simple analysis of the four observations in the model of price is a function of only organic and local certifications results in a measure of organic certification being worth 0.875 cents and locally produced being worth 0.625 cents on average. Conveniently, the example assumes that organic and local characteristics are independent or orthogonal and are measured as 0 or 1. These are common assumptions of simple regression techniques.

If this were a regression analysis, then the alpha would be the intercept and the beta would be the slope or sensitivity to the attributes or characteristics (organic or local).

This example of taking actual/revealed preferences data could be used to approximate measures of consumers' sustainable efficiency values (organic assumed to have better health benefits and better long-term environmental impacts) and consumers' social distance (here approximated

Table 2.2 Regression coefficients of local versus organic attributes

Price	Organic	Local
$ 5.00	1	1
$ 4.00	1	0
$ 3.50	0	0
$ 3.75	0	1
Alpha	beta organic	beta local
$ 3.31	$ 0.88	$ 0.63

by local). In a more complex analysis, buying local may also have benefits from reduced transportation costs (both monetarily and environmentally).

Arrow et al. (2013) in their debate on appropriate discount rates for governmental and in particular Environmental Protection Agency (EPA) decisions highlight the advantages and disadvantages of revealed versus stated preferences. There is also a theoretical underpinning to the choice of revealed versus stated preferences. The Stern (2007) report in its voluminous examination for the United Kingdom also provides guidance on appropriate discount rates for future generations within environmental or climate change decisions.

Portfolio Choice

So far, the discussion has been primarily about binary choices, A or B, and the transitivity of choices beyond only two, but still based on binary choices. The discussion has also focused on a single time period, choice made at time zero and held for one period with no changes. With a portfolio of choice, the discussion moves to combinations of different choices. Later, we will return to choices through time. As a starting point, the use of a food analogy and an investment portfolio will be employed. Consider the choice between vegetables and dessert for a child. The child (or at least mine) will prefer dessert over vegetables. However, after enough dessert, there will be a time that they will finally choose vegetables (we hope). Over time, one begins to understand the value of a balanced meal or a portfolio of foods. In the investment context, a model of Markowitz mean-variance portfolio, the optimal portfolio will be a combination of investments that either maximize return for a given level of risk or minimize risk for a given level of return, resulting in an efficient frontier of investment portfolios. The benefits of diversification become obvious in the investment context of risk and return. Portfolio choice represents better outcomes. The world is not binary, diversification of investments and even food make life better (Fig. 2.1). Issues of sustainability also benefit from diversification of choices.

Consumption Utility Model

Without bread (food) and water, humans cannot exist. The choice of food and liquids that are consumed is a consumption-based utility decision. A person will choose to consume within their budgetary constraints that which maximizes their utility. In the earliest part of human history, those choices of consumption were limited by opportunity and by budget. Even

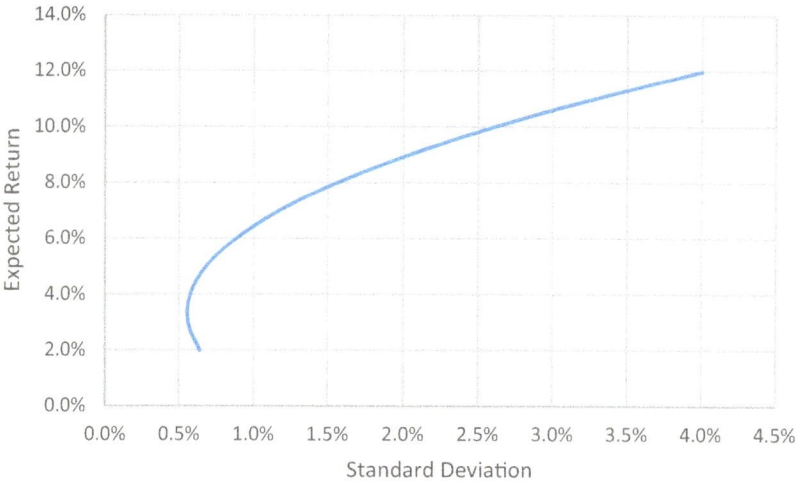

Fig. 2.1 Mean-variance efficient frontier

today, the difference in choice is one of economic availability through wealth and cost. Whether the discussion of the poor living on less than $2/day or the example of food deserts in lower-income neighborhoods of the United States, the choice of consumption is a choice driven by wealth and opportunity. In many economics' texts, the initial discussion may be of pizza and beer as a way of capturing the undergraduate imagination. A more recent example of pizza or hamburgers removes the ethical decision of underage drinking. Given issues of sustainability with a bit of humor added, the choice today may be pizza or meatless burgers. Regardless, economics is driven by resource scarcity and wealth with consumers making choices. This is also central to any discussion of finance and sustainability.

Even in terms of investment decisions, consumption choices are key. Think about different investments with respect to the time frame of their consumption goals. As a young adult, the first major purchase may be a car, then a house or condominium. Later, the big purchases may be children's education and finally retirement. In this light, each major consumption goal needs investment planning with an investment horizon and a level of acceptable risk. This implies different investment vehicles such as stocks, bonds, or alternative investments and an associated asset allocation among the investments. More sustainability investment choices are covered in Chapter 6 under the framework of different types of investors and their investment horizons and goals.

Much of the initial discussion of consumption choice is restricted to one-period models. The individual must choose today given their wealth and opportunities between two goods, pizza or meatless burgers (for goods x and y, in Fig. 2.2).

Returning to introductory economic texts reminds us that the optimal decision (maximization of utility) comes down to the marginal rate of substitution between the marginal utilities of the pizza, x, and meatless burger, y, and their respective prices and constrained by the consumer's budget line. The optimal choice given the budget constraint is the tangency of the indifference curve (a measure of utility) and the budget line.

Different individuals given their tastes (utility functions) and their wealth or budget constraints will make different consumption choices. Similarly, different individuals will make different choices about present and future consumption, thus having direct impact on sustainable or less sustainable consumption patterns. Individually, the reduction of today's

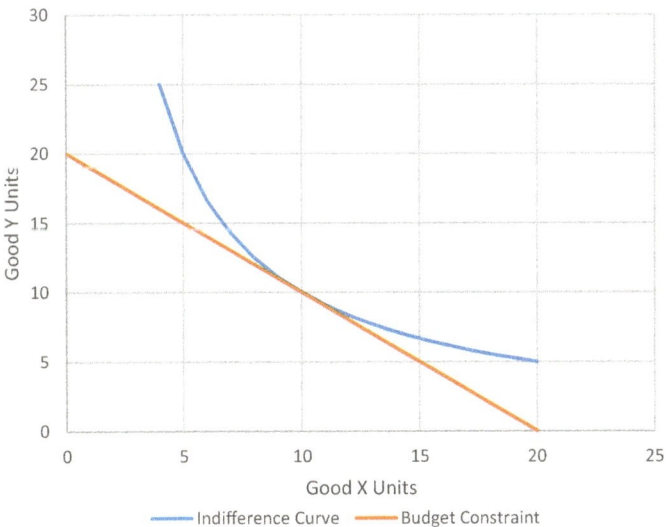

Fig. 2.2 Budget constrained utility optimal two product indifference curve

consumption and investment for future consumption will aid in more sustainable consumption patterns as well as reduced consumption. Societally as well, changes in consumption today will allow future generations to benefit in their consumption and thus meet one definition or goal of sustainability that of allowing the future generations to consume at similar levels as today's generation. It is thus imperative, as financial economists, to build a greater depth of understanding of the role of consumption-based utility models and their underlying assumptions.

Consumption Utility Model Through Time

Consumption models of utility lead us to the discussion of the time value of money and discounting. Consumers have the choice of consuming today (t = 0) or tomorrow (t = 1). The choice of consumption timing provides the discount rate or intertemporal time value or intertemporal marginal rate of substitution (IMRS), r. Another way of stating this is that a dollar today, P_0, is worth more than a dollar tomorrow, C_1, since the future cash flow is discounted by $1 + r$. As long as r is positive, this is true. The discount rate, r, is also referred to as the rate of return, the intertemporal marginal rate of substitution, yield to maturity, and cost of capital to name just a few of the terms employed in finance texts. Given a discount rate of r, a consumer should be indifferent between receiving P_0 at time 0 or C_1 at time 1.

$$P_0 = \frac{C_1}{(1+r)}$$

Typically, discount rates are assumed to be positive which resulting in "the dollar today is worth more than a dollar tomorrow" phrase for introducing the time value of money. This assumption may not always hold in the future, especially sustainable future with potentially lower economic growth. Given deflation in Japan, government rates have been negative. In the chapter on growth, the example of deflation or shrinking economic growth will be expanded. The implications of a shrinking population and more sustainable consumption patterns have important consequences to the type of economic growth observed through most of human history (see Piketty 2014).

For most finance texts, the starting point for an appropriate discount rate begins with risk-free rates matching the horizon of the future cash flow. US Treasuries fill this role from US Treasury bills with maturities of 0–1 year, Treasury notes, 1–10 years, and Treasury bonds, 10–30 years.

If there is no risk or uncertainty in future prices, then this is an appropriate assumption. The role of inflation or expected inflation in risk-free returns is best summarized by the Fisher equation where the risk-free or observed nominal rate of interest is approximately equal to expected real rates of return and expected inflation.

FISHER APPROXIMATION (1930)

$$R_f = Nominal \approx Real + Inflation$$

Once risk-free returns have been introduced, the next step is to move to an uncertain world. In an uncertain future, risk must be address. Common assumptions are that investors/consumers will be risk-averse. Risk aversion requires that investors be compensated for taking on risk, so the risky discount rates may generally be written as a risk-free rate plus a risk premium. If the risk premium is represented by an economic risk factor, then an asset's sensitivity to that risk factor, β, comes into the equation for a general asset pricing model (APM). Or as an example the Capital Asset Pricing Model (CAPM), where the risk factor is the market risk premium.

$$E(R_i) = R_f + \beta_i\big(E(R_m) - R_f\big)$$

SUSTAINABILITY UTILITY MODEL

Social distance measures allow the model to span decisions from corporate behavior through to philanthropic. It is easier to explain the model in reverse going from philanthropic to corporate. Given a base consumption model for the individual, to a certain extent the decision to be philanthropic expands the consumption model to the consumption of multiple individuals. The philanthropic decision is made on the basis of added utility to the donor from gifts to the recipient. The more efficient the philanthropy is in getting benefits to the desired stakeholders the more utility the recipient and thus donor should receive. In addition, the closer the social distance the more the reflective or "warm glow" utility the donor receives by giving.

This type of approach is reflected in the Bill & Melinda Gates Foundation's belief that "all lives have equal value," equivalent to a social distance of zero. *The Economist*'s "efficient altruism" is also consistent with this thinking. An easy metric for the efficiency of a charity is the proportion of funds that go toward mission. An efficiency of one would be 100% going to mission. Web sites are now available that translate the Internal Revenue Service 990 forms of charities into rankings of efficiency.

Moving on from charitable examples of social distance to other examples, there may be a relationship between social distance and physical distance. Politicians and environmentalists are more vocal in calling for people to buy local. The interpretation is that we know the local merchants and get added utility and added economic monetary benefits from buying local which is consistent with lower social distance and physical distance.

A simple example of social and physical distance overlapping would be buying local and organic produce. Assume you are examining the relative prices for organic broccoli to determine the health benefits value and social/physical distance value for the marginal broccoli purchaser with transitive preferences. Given broccoli from California (CA) costs $2.20/pound and organic broccoli from California organic costs $2.25/pound and organic broccoli from Rhode Island (RI) costs $2.40/pound. Assume the base price of broccoli is the California price. The consumer pays 5 cents more for organic and pays 15 cents more to get the organic broccoli from Rhode Island (the consumer is probably from New England).

Buy American or buy local highlights the role of social distance in consumption choices. An individual with low social distance to their neighbors might pay more for a local product than an international product. If the efficiency or quality were the same, then the price difference at which an individual was indifferent would reflect the social distance. If the social distance measure is modeled as a discount rate, then $U_{local} = \frac{U_{foreign}}{(1+\delta)}$ or the price of the local good is greater than the price of the foreign good. A "Buy American" choice would be how much cheaper does the foreign-produced goods have to be before the consumer switches. This will differ across consumers and businesses. Attitudes may change given marketing strategies or political agendas. Many governmental purchasing rules may have a preference for domestic purchases.

One of the more troubling issues with the environmental aspects of sustainability is the relationship to both current and future generations. The social distance parameter may be viewed as a substitute for very high discount rates for future generations that is future generations have a much greater distant and thus a much higher discount. If sustainability and environmental issues are couched in terms of the consumption ability of others (future and current generations), then decisions related to environmental impacts such as pollution reflect high social distances to others. Consumers who care less about the environmental impact would be those who care less about the impact of future generations because of a larger social distance (or their larger social distance is a result of their utility function having and attitudes to future generations). The reverse view or that of an environmentalist, from Chapter 1, would be the Iroquois notion of investing/decision making for seven generations out. Translation of the Iroquois social distance would be zero for all generations out to seven.

The concept of social distance(s) can be expanded for the business to different social distances to different stakeholders. In general finance, the shareholders would be viewed as having a zero social distance to the firm since they are the owners of the firm. The employees might be next in relative social distance followed by the customers, suppliers, and community. If the social distance is small enough, then it may be worthwhile approximating the effects on that group of stakeholders. The incentive movement in finance with employee stockholder options is to align employees and management with the owners by making them shareholders. The result for the sustainability consumption model is to shorten the social distance to zero for employee-owners.

LINE IN THE SAND OR A SPECTRUM OF OPTIONS

At the end of Chapter 1, there was a discussion of governments through laws and regulations setting down limits or lines of definition. Turning from the world of a "legal line in the sand" given legal definitions and regulations to a spectrum of options for sustainable business, the business organization turns from the minimum standard (legal or not) to decisions' impact on stakeholders. While rating or ranking systems often are also based on definitions of sustainable behavior and may exclude a firm from investments, they may also provide a spectrum of sustainable behavior through an overall rating system. A simple example is a rating system that uses screens as cutoffs. Sustainable indexes may restrict alcohol sales to less than 5% as an ethical screen. The five percent is the line in the sand. More of this will be covered in Chapter 4.

Returning to concepts of utility and value, when a hard line is drawn or strict definition adhered to the economic implications may seem absurd. The implication is that of an infinite cost to adherence. In the moment, decisions of absolutes may seem reasonable. In particular, the current political environment of populism is basically "my group before all others" or in social distance terms, zero social distance for my group, and an infinite social distance for anyone else.

Later in Chapter 6 within the discussion of risk, return, and sustainability trade-offs, it will become evident that how a business or investor handles such trade-offs will have a definite impact on the business and marketing of that business.

Binomial Pricing Models for Uncertainty

Another model common in financial and investment valuation is the binomial pricing model. Employing binomial models for sustainable decision making has many benefits for managers dealing with uncertainty. Binomial pricing models combine expected returns and their probabilities with discounted cash flows to model the dynamics of an uncertain future.

The mathematics of binomial pricing is straightforward. The probabilities of the two outcomes are the probability of one outcome, p, and one minus that probability of the other outcome, $1-p$ since the probabilities must sum to 100% or 1. The expected value is the probabilities times the value in that outcome state. In the equation below, p is the probability of the good or up (u for up) state and V_u is the value in the upstate. The downstate (D for down) probability is $(1-p)$ and its value is V_D. If those outcomes are in the future, then the addition of a discount rate, $1/(1 + r)$, brings the expected value from the future to an expected value in the present (Fig. 2.3).

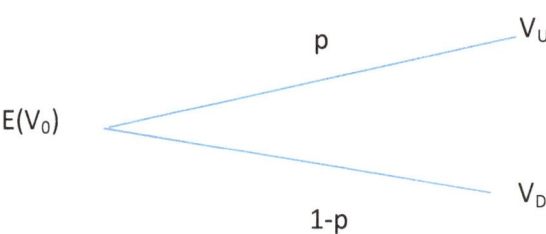

Fig. 2.3 Binomial pricing

$$E(V_0) = \left[\frac{pV_u + (1-p)V_D}{(1+r)}\right]$$

In a corporate decision making or investment model, the binomial pricing model may be useful in understanding the implied probabilities or valuations. Such a simple model is popular in mergers and acquisitions investing to find the implied probability of success or implied valuation of a successful deal. An investor may decide that the market's implied valuation or probabilities are incorrect and bet against the market.

An expanded binomial pricing model, that is one with multiple nodes and periods, is used for option pricing models to price American Puts and Calls with dividends. It could be used for interest rate and fixed income valuation for mortgage-backed securities with prepayment models as well. Given the simplicity of the two-node binomial model, its extension to more complex models and realities is that of only computational size and not necessarily of mathematical complexity.

For risk management and sustainability, one may understand the implied probabilities as that of an environmental disaster or social disaster. An analysis of corporate decisions pre- and post-disaster can be illuminative in the decision-making process, a number of examples old and new jump to mind.

- Did Ford understand and correctly price the probability of lawsuits and subsequent damages in their decision to build the Pinto in the 1970s without safety modifications?
- What is the change in disaster probability from oil tanker accidents such as the Exxon Valdez using double instead of single-hulled tankers?
- How has the public's and policy makers' probabilities and valuations changed with respect to nuclear power plant accidents after Three Mile Island, Chernobyl, and more recently Fukushima and the relationship to changing attitudes as a solution to climate change?
- The trade-offs inherent in the COVID 19 pandemic highlight the probabilities of death given certain types of economic and social closure.

A simple example may be illustrative of the power of binomial pricing models. Assume you are considering nuclear power plants as an approach to power that may reduce CO_2 emissions, but are concerned with the potential harm from spent fuel rods and a catastrophic accident. To avoid overcomplicating the analysis, the good state of the world is valued at

$100 billion of lower greenhouse gases (GHG), profits to utility shareholders and other stakeholders. The bad state of the world is valued, −$200 billion, from a loss of life and quality of life and lawsuits to the utility and its shareholders. For the moment assume both of these valuations are present value (already discounted with R and δ). The only unknown driving the decision is the probability, p, of the good state of the world, and thus also $(1-p)$ the probability of the bad state of the world. So the decision to build the nuclear plant goes ahead if the expected value is positive, $p(100) + (1-p)(-200) > 0$ or solving for the probability, build the plant as long as the probability of the good state occurs more than 2/3 of the time or the bad state less than 1/3 of the time. Any change to the probability of an accident going up or the cost of an accident going up would change the decision to build. Prior to the Fukushima nuclear power plant accident, there was more discussion of the benefits of nuclear power as a potential aid in reducing carbon emissions. After the accident talk of the benefits of nuclear power dissipated as new assessments of both the costs of a nuclear accident increased and the probability of an accident increased.

The translation of the binomial model to sustainability is in the uncertainty of the future and the probabilities of outcomes either positive or negative environmental or sustainable outcomes. Take the negative outcome example and apply the binomial model methodology to Ford's decision on the Pinto, Exxon's decision on the Exxon Valdez, Takata on the airbags, the Fukushima disaster, or any other cost-benefit analysis to take on a more sustainable production method. After the fact, it is possible to analyze the decisions these corporations made. For discount rates, historical expected returns from the CAPM may be used for each firm and the valuations before and after the incidents along with historical probabilities of such occurrences.

Edgeworth Box and Societal Agreement

Up until now, the discussion of utility has centered mostly on the individual economic agent's decisions. The individual may be viewed as a representative societal agent, generational agent, or shareholder representative. When decisions are made within groups in which there are many individuals, the conclusions reached in economic terms may relate more to game-theoretic conclusions. Such a discussion is beyond the scope of this book, but to provide context for similar decision-making processes in an economic utility format an introduction to Edgeworth box follows. Afterward, it will become clear why Friedman's "maximize shareholder wealth" is popular in the business community. It provides a simple path away from the messy decisions of consensus of disparate parties with disparate goals, social distances, and wealth. Similarly, the organizational decision structure around shareholders and stakeholders provides another path around consensus, more on this in Chapter 5. Sole proprietors may pursue their sustainable and profit agenda with less outside pressure than a corporation subjected to shareholder resolutions would. Like-minded partnerships, as long as they remain like-minded, would have a similar advantage in pursuing dual objectives. Consequently, B-Corporations with profit and goal objectives have tried to fill the gap.

When moving from one individual to two individuals and their respective utility functions, a useful framework is the Edgeworth box. In the context of an Edgeworth box, there are two agents with differing utility functions choosing two products (outcomes, goals) in a single-period model. Multiple period models of behavior between two agents fall under game theory which will not be addressed here.

The optimal decision for the two agents follows the tangencies of their respective indifference curves and the resulting Pareto optimal line connecting them. If it is not possible for the indifference curves to meet, then we have a basic political stalemate. This may be due to infinite costs for certain decisions by the two individuals.

Assume there are two agents trying to decide how to agree on the choices between profitability and sustainability. Three different cases should be sufficient to highlight the Pareto maximizing decision process First, if the two individuals have identical utility functions (very rare, but common among economic model assumptions), then they will agree on the decisions. The second case might be similar beliefs and utility that might bring them to agreement after finding mutually acceptable values and weights to their decisions. Finally, if one of the agents is only profit maximizing and has no interest (utility) in other outcomes, then a mutually beneficial decision may not be reached. Here the only profit maximizing agent has an infinite social distance to sustainable outcomes and the stakeholders impacted. To illustrate these cases, the following graphs represent Edgeworth boxes with the lower left and upper right be the origins of the two agents' utility functions with indifference curves emanating from them. Further from the origin represents greater wealth and utility. Tangencies between the agents' indifference curves represent optimal and agreeable decisions. The line connecting all the tangency points is called the Pareto set or contract curve (Fig. 2.4).

In Fig. 2.5, there is disagreement between the two economic agents in their utility functions with the resulting Pareto efficient curve or contract curve reflecting that agent one prefers y and agent two prefers x. The optimal point shows equal levels of wealth. Moving closer to either agent's origin would reflect less wealth relative to the other agent. If x and y were return or sustainability, the figure would reflect the imbalance in the agents' optimal portfolios, but would still represent optimal decision. Think of businesses with different social distances to shareholders and stakeholders coming to radically different portfolios of investment opportunities.

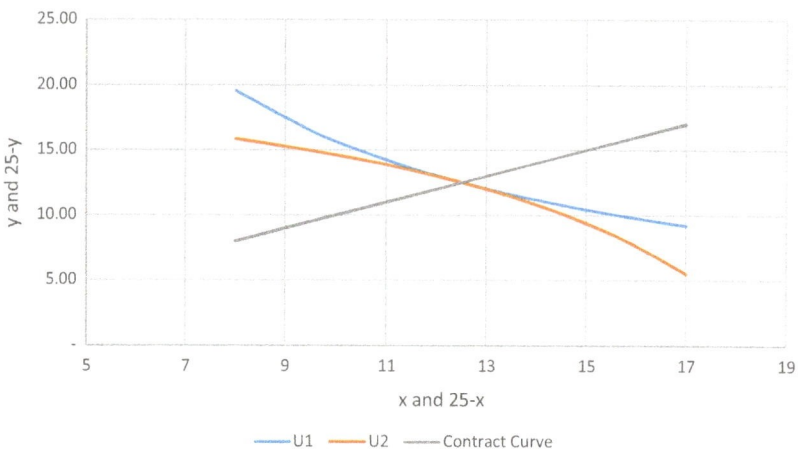

Cobb-Douglas Utility $u(x,y) = x^\alpha y^{1-\alpha}$ for agent one and $u(x,y) = x^\beta y^{1-\beta}$ for agent two, α and β = 0.5

Fig. 2.4 Edgeworth box similar utilities

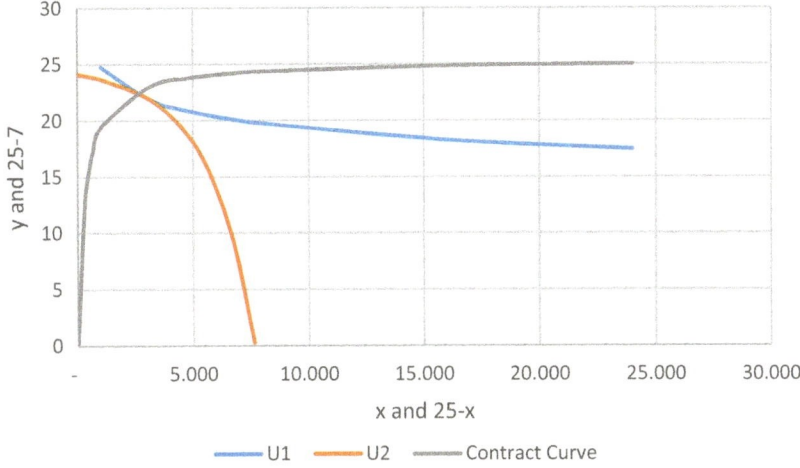

Cobb-Douglas Alpha = 0.1, Beta=0.9, x and y = 20

Fig. 2.5 Edgeworth box disparate utilities

INFINITY AND FINANCE

Infinite Costs

When thinking of the hard line in the sand analogy, the economist's response is that such a decision implies infinite costs. The decision maker is saying they will not change their position for any reason implies there is no amount of money that would change their mind. Ethical and moral decisions are often couched in a similar manner. They help to highlight the role of ethics in business decisions as well. It is these decisions that often are reflected in sustainability issues. The UN SDGs highlight others; those others are affected by business and governmental decisions. Public perception and opinion will often sway decision makers away from absolutes. Businesses need to be cognizant of their stakeholders in these instances, because missteps will impact the bottom line and shareholder wealth. The infinite social distance has been seen in 2020 in the growing "cancel culture" movement against corporations that do not meet the standards of certain groups. In terms of social distance, not caring about future generations or sustainability implies very high to infinite social distance. Discounting by infinity gives a value of zero.

Infinity and Going Concern

Another key occurence of infinity in finance is the going concern assumption in corporate finance assumes that a company lives forever. This assumption is translated into the discounted cash flow model as an infinite horizon for the cash flows. The Gordon Growth Model simplifies the infinite set of growing cash flows with a constant discount rate, r, into a growing perpetuity model. C_1 is the cash flow at time 1 and g is the constant rate of growth forever.

$$P_0 = \frac{C_1}{r-g} = \sum_{t=1}^{\infty} \frac{C_0(1+g)^t}{(1+r)^t}$$

Arguments for individuals and thus sole proprietorships and partnerships are extended beyond their lifetimes given inheritance which is the bequest motive to others. The bequest motive directly implies passing wealth and consumption possibilities to future generations. As a measure of social distance, the bequest motive suggests a shorter social distance to future generations. Giving future generations the opportunity to consume as current generations is also another definition of the stewardship aspect of sustainability.

In environmental economics, the problem with accounting for future generations is implied by huge future discount rates. Some have argued to not discount future generations as a means of forcing stewardship on models of sustainability. The social distance measure and efficiency measures are a different means to model and estimate current behaviors of stewardship. An organization or economic agent that does not consider future generations is modeled as having an infinite social distance to future generations or near-zero estimate of the efficiency of actions impacting future generations.

The concepts of infinity or an afterlife have implications for the economic agent. Pascal ventured into this philosophical and economic quagmire in the seventeenth century by positing that a rational person should act as if there is an afterlife (Pascal's Wager). If there is not, then there are only finite losses of lifetime consumption. If there is an afterlife, then there is the possibility of infinite gains for good behavior and infinite losses for bad. Mathematically, however, those infinite gains or losses are finite given our present value equations as long as the gains are finite and the discount rates positive.

With no discounting, an infinite set of cash flows will have infinite value.

$$\sum_{t=1}^{\infty} C_t = \infty$$

With discounting (even with positive growth, but with growth less than the discount rate, $r > g$), a growing cash flow is finite. Finance students will recognize this as the Gordon Growth Model.

$$\sum_{t=1}^{\infty} \frac{C_0(1+g)^t}{(1+r)^t} = \frac{C_0(1+g)}{r-g} < \infty$$

To delve momentarily into this religious and philosophical problem, there are tools that the economist has that may provide clarity or controversy. If "good behavior" toward others is sustainable behavior given the UN's SDGs and "good behavior" is enough to guarantee a positive afterlife if there is an afterlife given a certain probability (ignoring death bed conversions and their resulting religious beliefs of salvation), then the resulting analysis can be addressed using expectation resulting from assumed probabilities and present values given.

Just as the decision of infinite life or bequest motives have implications for sustainability, so does the social distance. A sustainable economic agent must have a shorter social distance to future generations and believe that sustainable actions have a higher value than a pure financial economic agent.

REFERENCES

Anderson, A., & Myers, D. H. (2018). Sustainability: Discounting the Future, Social Distance, and Efficiency Effects, *Moral Cents*, (Winter/Spring).

Arrow, K. J. et al. (2013). How Should Benefits and Costs Be Discounted in an Intergenerational Context? Working Paper Series 5613, Department of Economics, University of Sussex Business School.

Becker, G. (1968, March/April). Crime and Punishment: An Economic Approach. *Journal of Political Economy*, 76(2), 169–217.

Fisher, I. (1977) [1930]. *The Theory of Interest*. Philadelphia, PA: Porcupine Press.

Piketty, T. (2014). *Capital in the Twenty-First Century*. Cambridge, MA: Belknap Press.

Stern, N. (2007). *The Economics of Climate Change: The Stern Review*. U.K. Cabinet Office—HM Treasury.

CHAPTER 3

Growth and Business Sustainability

Abstract This chapter introduces a discounted cash flow approach to valuation of business ventures and highlights the key elements of growth. The success of a business, sustainable or not, is dependent on its growth. Expansion of a business relies on expanding its markets' reach (demographics), its innovation or creation of new products, and its capital structure (leverage) within a legal and governmental framework. The interplay of social distance and demographics highlights the markets that a business may expand as it builds customer relationships. Those relationships will be based on the marketing of innovative sustainable products to new markets. The role of leverage on capital structure will be laid out within a risk and return framework.

Keywords Discounted cash flow · Sustainability · NPV · Social distance

The success of a business, sustainable or not, is dependent on its growth. Expansion of a business relies on expanding its markets' reach (demographics), its innovation or creation of new products or more efficient processes, and its capital structure (leverage) and the legal or governmental structure to which it is subject. The interplay of social distance and demographics highlights the markets that a business may expand as it builds customer relationships. Those relationships will be based on the marketing of innovative sustainable products to new markets. The role of leverage on capital structure will be laid out within a risk and return framework.

Do not forget that financial sustainability requires positive expected wealth creation. Financial profit is necessary for survival. The aim becomes how to create sustainability within existing organizations or by creating new organizations. It is with deliberate purpose that the term organizations is used. While it is straightforward to think along the line of business or corporation, sustainability is important along all organizational structures. The creation of dual-purpose corporations, such as B-Corporations, also reminds the student or researcher that non-profits, sole proprietorships, non-governmental organizations (NGOs), religious organizations, educational organizations, and other structures must balance sustainability and financial health as well. This will be covered in more depth in Chapter 5.

In the simple model of valuation, the discounted cash flow model, the key variables are cash flows and the discount rate through time. Chapter 2 discussed the role of social distance on the observed discount rate. The cash flows for a healthy business grow through time. The Gordon Growth Model assumes that there is both a constant growth rate, *g*, and a constant discount rate, *r*. Think back to last chapter's discussion of infinite horizons.

$$P_0 = \frac{C_1}{r-g} = \sum_{t=1}^{\infty} \frac{C_0(1+g)^t}{(1+r)^t}$$

Growth comes from three main areas for a firm (and for the economy)—leverage, innovation, and demographics. This chapter discusses the role of each within the context of a sustainable business.

The DuPont Identity is a popular method in finance to disaggregate growth. The DuPont Identity is dependent upon the disaggregation of the return on equity for a firm.

Given that growth for a firm depends on the money a firm invests of its own earnings (the plowback or retention ratio, b, is the percent of earnings retained for future growth) and the return the firm gets from its investment (ROE). Growth, g, may summarized as

$$g = b * ROE$$

Return on equity (ROE) = NI/TE = PM*TAT*EM = (NI/Revenue) (Revenue/Assets) (Assets/Equity)

- NI is net income
- TE is total equity
- PM is profit margin
- TAT is total asset turnover
- EM is equity multiplier

The equity multiplier leads to the first driver of growth, leverage. Since total assets include total equity and total debt, then the ratio of assets to equity provides an indicator for the level of debt (leverage) that a firm has.

Leverage

Leverage comes from borrowing. It magnifies returns on both the upside and the downside, thus increasing risk. Individuals through credit cards, car loans, home loans, or even student loans increase their leverage through their consumption. Businesses borrow from banks, financial institutions, and financial markets through loans and other debt instruments. Governments borrow through the issuance of government bonds, such as municipal bonds or Treasuries in the United States or sovereign bonds from other governments. The Great Financial Crisis saw a large decrease in private borrowing in mortgages to an increase in governmental borrowing. This shift from private to public borrowing also highlights the shift from this current generation to future generations. The tradeoffs between generations parallels the argument that sustainability represents the emphasis on impacts on current and future generations' ability to live (or consume) in a sustainable manner.

A quick example of the power of leverage to increase risk and return is borrowing to buy a home. For example, an individual could borrow $100,000 to buy a $500,000 home (this is the blue line with 20% borrowing and borrow $250,000 or 50% is the orange line). If the value of the home rises by or falls by $100,000 or 20%, then the resulting returns on the sale of the house are gains of 25 and 40%.

Initial wealth of $400,000 with a loan of $100,000 buys a $500,000 home. The value rises 10% to $600,000 and sold (assuming instantaneous price change so no interest on the loan). Payoff the $100,000 leaves $500,000 and a return of $100,000/$400,000 or 25% (Fig. 3.1).

The leverage from borrowing increased returns and losses from 10 to 50%. It should be remembered that it also allowed the individual with

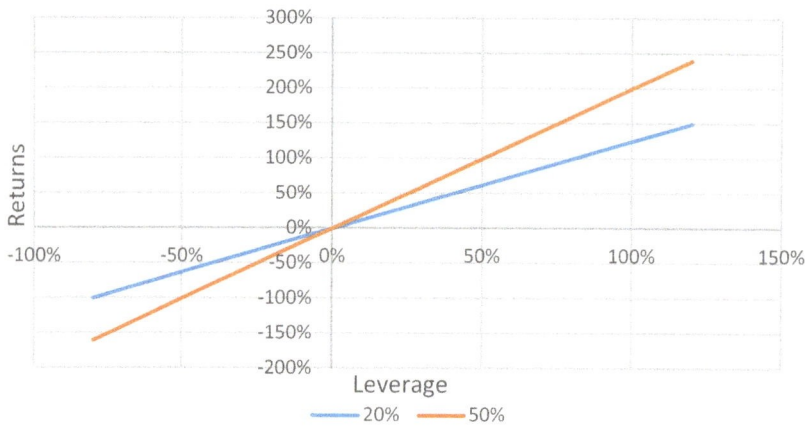

Fig. 3.1 Impact of leverage on returns

only $100,000 to invest to buy a $500,000 home. In a similar manner, businesses (sustainable for not) may use leverage to grow. Obviously given the increased risk, businesses must understand the financial sustainability of their decisions.

Two fundamental articles in corporate finance are by Modigliani and Miller. The papers highlight the role of debt (leverage) in the capital structure of the firm. In the first article, MM1, for which they won the "Swedish Financial Prize in Economics for the Nobel," they argued that given that individuals may mimic the leverage of the firm, then leverage should be irrelevant to the value of the firm. The second article, MM2, expanded the analysis to include the benefits of the tax shield on debt weighed against the costs or probability of bankruptcy. Given the trade-offs of the costs and benefits, MM2 shows that there must be an optimal level of debt (or leverage) for the firm (see Fig. 3.2).

The relationship of leverage and optimal capital structure in the realm of sustainability arises in the observation that better run companies may have lower risk which would afford them lower costs of capital (equity or debt).

$$WACC = w_e k_e + (1 - \tau) w_d k_d$$

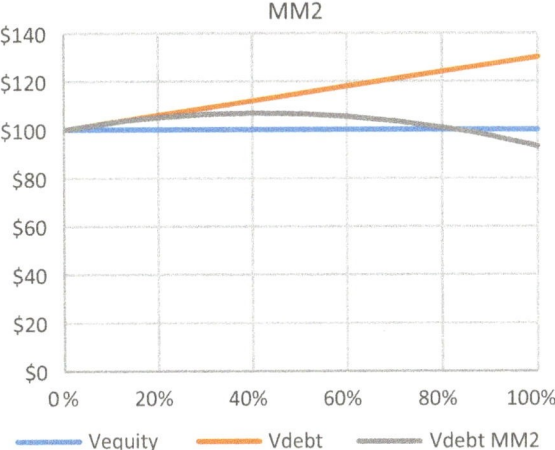

Fig. 3.2 Leverage, capital structure, and bankruptcy

Where the weighted average cost of capital (WACC) is the weights in equity and debt and cost of equity and after-tax cost of debt. Lower costs of capital improve growth opportunities by providing more positive net present value projects. So, if a firm is more sustainable and that brings about lower risk from avoiding bad outcomes (unethical behavior, oil spills, lawsuits …), then the firm may be able to lower its cost of capital and be able to afford more new projects for more growth.

Innovation

Innovation comes in many different forms. For the broadest definition, innovation is a new product or process. Much of the growth in sustainable businesses will come from innovations. The efficiencies in alternative energy creation and storage represent the most relevant and current examples in innovation. Lower costs from manufacturing efficiencies, increased production capabilities, and increase in market (see demographics below) have helped grow the solar industry. The growth was also aided early on by government incentives for the solar industry that brought the marginal costs of production down with the growth of the industry which drove further growth.

Returning to the DuPont identity for a moment, it may be possible to use profit margin or total asset turnover as indirect measures of innovation or efficiency. The more efficient a process becomes the higher the profit margin (Net Income/Revenue) should be or the higher the total asset turnover (Revenue/Total Assets).

Demographics

Demographics for a business are the customers and consequently the sales or revenue. The more customers or richer customers you reach, the more revenue is created. The decisions surrounding which customers to reach are typically decisions of particular demographics the business is targeting. The customers may be defined demographically by population, gender, age, wealth, religion, ethnic background, location, or other criteria. Marketing and sales targeting are aimed at capturing a greater or wealthier set of customers. It is in this aspect that demographics play a very important role in business and thus in business growth. In terms of social distance, marketing to a certain demographic is reducing the social distance to that demographic or set of stakeholders.

Returning again to the DuPont identity, total asset turnover (Revenues to Assets) may also be interpreted as the demographic reach and its success for the firm. Generating more sales through more or wealthier customers for a given level of assets is demographic success.

Recent global trends in economic growth or lack thereof highlight the reason for the ascendancy of Chinese and the fall of the Japanese economies. While Japan's growth during 1980–89 may be attributed to leverage and innovation, the stagnancy of the economy since 1990 has been exacerbated by the aging population and low birth rates. China on the other hand saw their economic growth come from the growth in their middle class which was aided by public and private leverage. The growth in the middle class is reflected by the movement from rural areas to the cities and by growth in literacy rates. This was accomplished even with the one-child policy—a policy that has been abandoned to support future aging populations' retirement. An obvious nod to avoid the Japanese fall. In the United States, there has been lower immigration and a shrinking of the middle class both have a dampening effect on the economy.

These examples also highlight a conflict between economic growth and sustainability. Even within the United Nations Sustainable Development Goals, the trade-offs of better health, education, and economic equality between developed and developing economies put a strain on environmental issues and climate change. The growth of the Chinese economy and the resulting growth in greenhouse gasses from its rapid growth have put a greater strain on the environment—a pattern that is likely to reoccur with India and with other developing economies. This appears to be in conflict with the environmental Kuznets curve (EKC) and its projections of reduced environmental impact from advancing economies. The EKC may work on the local level, but does not appear to hold globally as developing economies try to catch up. The growing economies of the developing world are creating the same environmental impacts of the developed world.

Demographic issues and economic growth raise other conflicts within sustainability. The past half-century since the end of the baby boom (early 1960s) has seen efforts to reduce population growth as a means of sustainability. The zero population growth movement was most successfully implemented through China's one-child policy (a policy that has been abandoned for economic reasons). This tension has been abundantly clear through sustainability and economic growth in most demographic and economic growth scenarios. China's recent economic growth came from the growth of the middle class and their newfound wealth and power. The result environmentally has been to add other huge economies producing greenhouse gases. More people or richer people mean more consumption which leads to more pressure on the environment. Development in economic strength is correlated with lower birth rates (statistics of most OECD country birth rates are dramatically lower than African and other developing nations). The economist sees the monetary substitution from economic growth replacing the agrarian large family model. This is the same tension in the UNSDGs between the health and welfare issues and the environmental issues.

Returning to the individual and the sustainable business, the role of demographics and capturing a demographic is one of sales and marketing and changes of consumption behavior. Innovation assists in the changing behavior by lowering costs and increasing efficiencies of products. In studies of generational differences in measuring the importance of sustainability choices, it is already evident that the millennials are more interested in the growth of sustainability in businesses than older generations.

The discussion of growth in general begs the question of whether the economic growth seen in the past two hundred years since 1800 is sustainable. If not, what are the political ramifications? Reduced birth rates and aging populations have had a dramatic impact on the Japanese economy, and fears of similar slowdowns are evident in China's reversal of the one-child policy. These concerns lead directly into less a business' role than a governmental or societal role in growth. These concerns about population are less relevant to a business than they are to governments and society about appropriate levels of growth.

Adapting the Gordon Growth Model to social distance by introducing δ gives the model the form of

$$P_0 = \frac{C_0(1+g)}{(r+\partial)-g}$$

Thus, an increasing social distance, δ, provides lower value and a near-zero social distance gives greater value to the stakeholders that the social distance is in relation. Growth is still a very powerful motivator in valuation with or without social distance.

ROLE OF GOVERNMENT

The role of government and its effects on the growth of companies comes in many different forms from laws, regulation, enforcement, subsidies, and initiatives. Even in the midst of the Great Financial Crisis, there was positive movement in green stocks in anticipation of the new administration. The markets believed that government's attitude in all its forms would favor green companies. Globally, efforts by governments to move toward renewable energy have changed the cost structure of generation as marginal costs of production have fallen. China and Germany through production and demand, respectively, have moved the solar power industry forward. The reduction in solar energy production costs has also been aided by technological advances or innovation. Over the Christmas weekend of 2017, energy prices were reported to be negative in Germany due to the overproduction of electricity from renewable sources and lower demand from mild weather.

Growth is key to all businesses and their valuations. These three drivers—leverage, innovation, and demographics—have different impacts on the sustainability of a business as well. Innovation may be a driver for both growth and sustainability if the innovation's direction is towards more sustainable products and processes. Demographics inherently have a sustainable or social justice angle. As a business expands its demographic reach, it may decide to serve underserved populations while still provide opportunities for growth. All of this growth requires the legal and regulatory structure to nurture it.

CHAPTER 4

Certification of Sustainability

Abstract This chapter introduces the reader to different ratings created and employed to measure sustainability. In building relationships with consumers for sustainable products or business, the trust and certification of sustainability is dependent upon measurement. The chapter will discuss methods of certification through organizations such as LEED, Bloomberg's ESG, MSCI KLD, and other assurances of sustainability. The strengths and weaknesses of the measurability will be highlighted by the presence of false advertising known as "greenwashing."

Keywords ESG ratings · Sustainability · Social distance · Business

In building relationships with consumers for sustainable products or business, the trust and certification of sustainability is dependent upon measurement. This chapter will discuss methods of certification through organizations such as LEED, Bloomberg's ESG, MSCI KLD, SASB, Fair Trade, Organic, and other assurances of sustainability. The strengths and weaknesses of the measurability will be highlighted by the presence of false advertising known as "greenwashing."

The key to a good certification system is trust between the consumer and the business and also the ratings system or certification. From the trust will come acceptance. With acceptance will come more trust and the cycle will continue. Acceptance and trust often begin with a first-mover advantage. A great financial example of this first-mover advantage is the Dow Jones Industrial Average, a US stock index of 30 stocks, and the S&P 500 index. The Dow still receives the first headline of the day and the S&P 500 gets the next. The Dow started in 1896 is more widely known and accepted, but is an imperfect measure of stocks as it is a price-weighted average limited to only 30 stocks. The S&P indices were the first and more accepted market value-weighted indices. Market value-weighted indices are better reflections of potential investment opportunities and more accepted for performance measurement. The progression of equity indices reflects first-mover advantage. For example, international equity, the Capital International EAFE Index began in 1969, now the MSCI EAFE index, for small capitalization US stocks represented by the Russell 2000 began at the end of the 1970s. The popularity and acceptance of these indices is reflected in the volume of futures and options traded as a measure of trust. Across all of these examples, there is a large premium to being first. Even in sustainability certifications being first may be the key to success (LEED, Fair Trade, et al.).

A key caveat to rankings and measurement is to remember that "everything is measured with error" from the earlier discussion of Plato's Allegory of the Cave. The importance is that perfection in measurement is neither a goal nor a possibility. Certification and measurement need to minimize the errors in order to increase the believability of and trust in the process. One of the main observations of the growth in sustainability is that as there has been a growth in certifications, there has been a growth in usage. A simple economic explanation of these phenomena is that as the marginal cost of measuring (even with error in measurement) has decreased, there has been a resulting increase in adoption. As the measurement gains acceptance in general, the measures may also be improved upon (less error) and lower marginal cost of measurement will grow the overall acceptance.

Trust and Acceptance in the Certification

For any certification program to work, there must be trust in the system and widespread acceptance. Consumers who trust the impact or relevance of say fair-trade certification would be more willing to pay more for products with the certification than for products without. The value of the certification to the consumer may be measured in terms of revealed preferences through observed purchases. In the era of big data, the point of sales information collected by a retail business would provide a great starting point. Counter to this and probably the biggest hurdle in sustainability is the issue of "greenwashing" in which a firm spends more money marketing the perception of environmental savings than the actual savings. If consumers doubt or mistrust a firm's claim or a rating systems ability to measure the environmental benefits or sustainable benefits than the exercise in sustainability will fail. Consumers must be confident that they are getting what they are paying for in more sustainable products or processes.

Assume that a consumer (or many consumers) has the choice between certified fair-trade coffee and non-certified fair-trade coffee. If across all other attributes the two coffees are identical (packaging, taste, aroma, ...) and thus the only difference was the fair-trade certification, then the price difference would be a reflection of the value of the certification. Since it is rare for two products to differ by exactly one attribute, the big data approximation would be to run regressions of coffee prices and attributes across thousands (or more) observations to approximate the value of the certification.

Price of fair-trade coffee = Value of certification + Value of other attributes

Price of coffee without fair trade = 0 (value of certification) + Value of other attributes

Price of fair-trade coffee—Price of coffee without fair trade = Value of certification if the Value of the other attributes is identical.

Measurability of the Certification

With any ranking or certification program, there will be measurement issues. Think of the college ranking systems done by the likes of the Princeton Review and US News & World Report for universities and colleges in general or Bloomberg, Poets & Quants or others for business schools. The weights of the factors used to rank the schools and the factors will differ across the provider. The measurement within the factors will be measured with error and the errors may be dependent on the provider/measurer's biases. The certification will gain in popularity and trust through time with consistent and transparent measurement. With trust and popularity, the certification organization may be able to charge the firms producing the products a higher price based on the wider acceptance and proven value of the certification.

Relationship of Certification to Social Distance

Returning to sustainability ratings, who the beneficiaries or stakeholders and the social distance to them are relevant to a certification program. For Fair Trade, the beneficiaries, those with short social distance, are farmers and farmworkers. For organic products, implying low to no pesticides, current generation consumers as well as future generation consumers benefit that is positive non-zero value. The LEED certification for buildings also aims to benefit current and future generations. ESG ratings help investors match their social distances and morals to investment vehicles. The Carbon Disclosure Project highlights the environmental issues of global warming for investors. Two hypothetical examples are addressed below (Tables 4.1 and 4.2). In setting up a rating system it is necessary to implement cutoffs for passing certain criteria or levels of criteria and then to assign weights to the criteria (remembering that the weights must sum to 100% or 1.

In creating a rating system, the criteria, scores, and weights must be laid out. The analysis of the products and measurement against the criteria is made. A final score based on the measurements, criteria, and weights is then decided. The tin can with a string in Table 4.2 passes the sustainability test given the weights and cutoffs assigned in Table 4.1 if it is necessary to achieve a ranking of 3 or above. Similarly in the investment ranking or rating systems, as Berg et al. (2019) discovered with ESG ratings, there are errors and differences across these dimensions of weights and criteria measured. Those errors and differences in criteria and weights explain how different investments and portfolios receive widely different ESG rankings.

Relationship of Certification to Stakeholders

Sustainable certification measures must be related to benefits for stakeholders. Sustainable food products should have health benefits or environmental benefits. The health benefits should be categorized as a benefit to consumers; the environmental benefits should be better categorized as having benefits to current and future generations and the increase in value of those benefits are indicated by having lower social distance or discount

Table 4.1 Rating criteria example

	Sustainable cutoff	Sustainable cutoff	Sustainable cutoff	Sustainable cutoff	Sustainable cutoff	Sustainable cutoff	Weight (%)
Points	5	4	3	2	1	0	
Wages	120%	115%	105%	100%	90%	below 90	10
Work conditions	35	40	45	50	55	60	5
Toxic ingredients	10%	20%	30%	40%	50%	60%	30
Recyclable	100%	80%	60%	40%	20%	below 20	30
Accessibility	100%	80%	60%	40%	20%	below 20	25
Level of sustainability	5 stars	4 stars	3 stars	2 stars	1 star		
Total score	100	80	60	50	40		100%

Table 4.2 Rating examples

	Smart Phone	Cell Phone	Tin Can as a Phone
Wages	4	3	1
Work conditions	2	3	4
Toxic ingredients	1	2	4
Recyclable	3	4	5
Accessibility	4	1	5
Total score	54	50	85
Sustainable	No	No	Yes

factor. The consumer who buys an electric car finds benefit and value through being more efficient in terms of energy usage and having a lower social distance to current and future generations given lower emission of greenhouse gases (GHGs). Similarly, LED lighting systems provide more value to the consumer through both efficient energy usage and also lower social distance to other generations.

SUSTAINABLE ACCOUNTING STANDARDS BOARD (SASB)

In terms of compliance with standards for businesses in sustainability, the US lags other parts of the world. One organization attempting to change that is the Sustainable Accounting Standards Board (SASB). SASB has a certification system for "61 industries." There is movement to get the Securities and Exchange Commission to adopt some of their recommendations. Sustainability reporting in the United States is mostly voluntary. Oceanian and European countries are further along on requirements for sustainable reporting. Reporting is still couched in terms of ratings and break points. Business adherence is thus in transition. Businesses can still choose to be ahead of the regulatory curve for a competitive advantage.

GOVERNMENTAL ROLE IN CERTIFICATION

There are national and global standards for certification of sustainability measures. These include requirements for disclosure and reporting as well as meeting certain standards to meet definitions of sustainability. Some countries and regions are further along the path of sustainable reporting than others. Leaders have been the European Union, Australia, and New Zealand.

The International Standards Organizations (ISO) which set out guidelines for companies to follow across many industrial categories have sustainability standards as well. Their global standards include ISO 14000 for Environmental Management, ISO 9000 for quality and efficiency, and ISO 26000 for sustainability.

Stability and Consistency

For the certification to be trusted, the characteristics need to be consistently applied through time. The consistency comes from the stability of rankings through time. Changes to rankings or certification must be clear and transparent to the users. Having pointed out the difficulties of the line in the sand analogy and the implied infinite costs, human behavior seems to want to know the boundaries. Thus, a good certification program will delineate their criteria even when it will come up against seemingly arbitrary results. In 2006, Starbucks had more than 5% of its revenue come from global sales related to products with alcohol. The 5% represented a threshold for some socially responsible investors as a cutoff for alcohol-related investing. There was some outcry in the Social Responsible Investment (SRI) industry about whether the 5% limit was arbitrary and that the firm did well in other categories to warrant its inclusion. This is a clear example of the problems with cutoffs in rankings and disagreement on the weights given to different criteria.

In a social distance example, those arguing for buying domestically run up against what is a domestically produced product. If the criteria were 100% domestically sourced and produced, it would be a very high, if not impossible, hurdle to clear. Businesses are aware of these issues as they move production facilities around the global to avoid such issues.

In creating a certification/ranking system for a product or process, it is necessary to think about the criteria, the measurement, and the weights for each criterion. Berg et al. (2019) found that differences in these three areas drive most of the differences among ESG ratings of corporations. If one were to create ratings tied to the United Nations sustainability goals, Table 4.3 could provide a starting point. The categories have been set, but there may be individual criteria within each of the 17 categories that would still need to be ranked. This is the score column for a fictious

Table 4.3 UN SDGs as a rating system for an organization

	UN SDGs	Weight (%)	Score
1	No poverty	5	4
2	Zero hunger	5	5
3	Good health and well-being	8	2
4	Quality education	9	3
5	Gender equality	5	4
6	Clean water and sanitation	8	5
7	Affordable and clean energy	8	6
8	Decent work and economic growth	8	2
9	Industry, innovation, and infrastructure	4	2
10	Reduced inequalities	5	2
11	Sustainable cities and communities	5	2
12	Responsible consumption and production	5	2
13	Climate action	5	2
14	Life below water	5	2
15	Life on land	5	2
16	Peace, justice, and strong institutions	5	2
17	Partnerships for the goals	5	2
		100	3

organization. In terms of ranking, at a minimum, there should be a preference ranking (think back to Chapter 2 and transitive preferences and how those choices translated to the rank ordering of the UNSDGs). Even if the criteria can be measured, it is typical to have added cutoffs so that a differentiation in scores may be made. In combination with the individual organization scores on each criteria or SDG, in this case, there must be an overall weighting system (the weight column in Table 4.3) that is applied consistently across organizations.

The weights, criteria, and measurement will all be impacted by the rater's utility function and implied social distances to the people impacted by the criteria. The difficulty in the measurement is not trivial and the interpretation of Plato's Allegory of the Cave reminds one that everything is measured with error. But the process of creating and reviewing the rating system will be based on a specific time in which there should be a preference ranking, preferred or indifferent.

The Excel functions in the bottom two cells are $= \mathrm{sum}(D4{:}D20)$ and $= \mathrm{sumproduct}(D4{:}D20, E4{:}E20)$ which are equivalent to $\sum_{i=1}^{17} w_i$ and $\sum_{i=1}^{17} w_i r_i$

Remember that the weights, w, must sum to 1 or 100%, and r's are the ratings. It is not necessary for each of the 17 SDGs to be employed and for some products, sub-categories, may be appropriate and then aggregated for the SDG score. The production of each component of a product may be scored to reach an aggregate score for the product or the firm.

Finally, whether as a business employing a rating system to validate a product or as a rating system certifying products the benefit of first-mover advantage for rating systems cannot be understated in gaining trust with consumers. With clear methodology, consistent criteria and measurement combined with being the first in the field will provide a business opportunity for more areas in sustainability products and processes.

Reference

Berg, F., Koelbel, J. F., & Rigobon, R. (2019). Aggregate Confusion: The Divergence of ESG Ratings, MIT Sloan School Working Paper 5822–19.

CHAPTER 5

Corporate Implementation and Business Forms

Abstract This chapter highlights the different legal structures for business and how sustainability fits with those structures. Corporate or business implementation strategies are covered. Implementation of structure and process in a sustainable framework highlights investments in products through a sustainable net present value (SNPV) process or business structure. Differences in structure and form such as B-Corps, non-profits, institutions, partnerships, and corporations are discussed within the area of sustainability.

Keywords NPV · Sustainable NPV · Business form

When a business decides on whether or not to invest in a particular product, machine, or project, the business must have a straightforward and consistent methodology of choosing among investments. Those decision criteria include net present value (NPV), payback and discounted payback, simple and straightforward, and Internal Rate of Return (IRR). Most financial texts will highlight these methods and others with the general conclusion that NPV is the preferred method. For a sustainable business or organization, NPV is still the best choice. In addition, given the view of stakeholders, a derivative of NPV will be introduced here, a sustainable NPV (SNPV). SNPV will be defined as including the related advantages and disadvantages to other stakeholders consistent with social

© The Author(s) 2020
D. H. Myers, *Sustainability in Business*,
https://doi.org/10.1007/978-3-319-96604-5_5

distance and discounting. In truth, if social distance is never infinite, then current decisions are still based on SNPV > 0.

Given that other valuation methods are employed in practice, this chapter highlights some of their shortfalls in general and with respect to sustainable implementation strategies. In addition, the business or organizational form and structure are important to the process of the implementation of investments in products within the sustainable business structure. The differences in structure and form such as B-Corps, non-profits, institutions, partnerships, and corporations may shape the issues of both profitability and sustainability (the trade-off of maximizing shareholder and stakeholder value).

Net Present Value and Discounted Cash Flows

As a starting point for decision making, a core financial approach is presented, net present value. Net present value is one of many derivations in financial economics of the discounted cash flow method. The most popular phrase for discounted cash flows is *that a dollar today is worth more than a dollar tomorrow*. The simplest translation is that if you had a dollar today and could invest it in a risk-free investment for one period, then you would have more than a dollar at the end of the period. While this is the norm for beginning discounted cash flow analysis, the world's economic growth rates have slowed. So, a dollar invested today, $\$1.00(1+R_f) > \$1$, will be worth more as long as the risk-free return, R_f, is positive. The twenty-first century has seen nominal negative interest rates in Japan, Germany, and Switzerland and elsewhere. For most of the analysis that follows, negative interest rates will be ignored. In terms of long-run analysis of economic growth in a more sustainable world, negative nominal interest rates may become more prevalent. One way around that long-term problem of negative interest rates is to couch investments in positive real returns.

To begin the discussion of discounting, start with a single cash flow and a single investment period. For whatever cash flow or investment you have today, C_0, it may grow by $(1+R_f)$ in the future to a value of C_1. Discounting is just a rearrangement of this relationship between a future value, C_T; a present value, C_0; and some return or discount rate, R, over T periods.

$$C_0 = \frac{C_T}{(1+R)^T} = \frac{1,000,000}{(1+.04)^{20}} = 456,386.95$$

$$C_0 = \frac{C_T}{(1+R)^T} = \frac{1,000,000}{(1+.04073)^{20}} = 450,000.00$$

$$C_0(1+R)^T = C_T = 450,000(1+.04)^{20} = 986,005.41$$

For example above there are three approaches to consider the variables in a discounted cash flow—present value, future value, and the discount rate. If you were to receive $1 million in 20 years or $450,000 today and invest the money today at 4% per year, then the $1 million would be equivalent to $456,386.95 and would be the better option. Given the manipulations that are inherent in the approach and discussion of valuation, one might also consider that the $450,000 would grow to only $986,005.41. Or finally, the investor would be indifferent if the discount rate was approximately 4.073% per year yielding the $1 million in 20 years. Each answers the same problem, but solves for a different variable—C_0, C_T, and R. When solving for R, one finds the internal rate of return (IRR).

If an investment has many future cash flows, C_t, then the value today of those future cash flows is just the sum of all the discounted cash flows. To keep the analysis simple, the assumption of a single discount rate, R, is made, and for the moment, all cash flows are known.

$$C_0 = \sum_{t=1}^{T} \frac{C_t}{(1+R)^t} = \frac{50}{(1.05)} + \frac{100}{(1.05)^2} + \frac{250}{(1.05)^3} = 354.28$$

A business decision maker or an investor wants to ensure that an investment pays off or adds value for themselves or their shareholders given the relevant level of risk. The investment or project may have numerous future cash flows, C_t, as well as the initial outlay, C_0.

Rearranging the equation to the typical NPV form highlights that NPV informs the decision maker if the future discounted cash flows exceed the investment costs given the appropriate discount rate. If NPV is positive, then the project should add value to the shareholders above the appropriate discount rate, R. The NPV rule accepts the project or investment, if NPV is positive.

$$\text{NPV} \geq -C_0 + \sum_{t=1}^{T} \frac{C_t}{(1+R)^t}$$

where the basic inputs are the net cash flows per period, C_t, and a discount rate, R. For simplicity, the assumption is for the discount rate to be constant through time. For firm valuation, the life of a firm is assumed to be infinite. This is known as the "going concern" assumption. For a project or investment, there would typically be a maturity, T. The NPV for an investment is the initial outlay, C_0, plus the net cash flows through time discounted. The investment rule is to accept projects with positive NPVs; that is, the return from future cash flows exceeds the initial outlay. In comparing two projects, choose the project with the higher NPV given no funding restrictions. Obviously, if there are insufficient funds available, the projects are not possible.

To be able to implement the NPV rule, the correct or proper calculation of the inputs is necessary. Cash flows must be related to the investment such as marginal costs of taking on the project or the opportunity costs that the project represents. The idea of opportunity costs or positive and negative externalities highlights the need in both traditional and sustainable decision making to account for ALL THE RELEVANT CASH FLOWS. The emphasis on relevant and incremental cash flows, C_t, is particularly important in distinguishing NPV (shareholders) and SNPV (shareholders and other stakeholders). If there are benefits to increased sales from improved sustainability marketing efforts, then those cash flows should be credited to traditional NPV inputs. Financial managers, as step one in sustainability, must account for all the positive and negative effects correctly in NPV even before moving to an SNPV. Examples of shortcomings include Ford Motor and its decision to build the Pinto without proper safety changes in the 1970s or Boeing's 737 Max rollout and Volkswagen's diesel emissions debacle in recent years. They may have come to a different conclusion if the impact (both cost and probability) had been correctly modeled. This analysis may couched in terms of binomial pricing models to capture the probabilities and value impacts on lose of life or quality of life to stakeholders.

After the cash flows have been correctly modeled, the next step is to appropriately model the discount rate. A business' discount rate is related to the risk of the business and its projects. Traditional finance relates the risk of the firm to the capital structure or leverage, the business or industry in which it operates. The industry will give an indication of the typical business risks related to operating leverage in the industry and the cyclicality of the revenues. These factors will affect an organization's cost of capital for the cost of the funds employed for a project. The firm's cost of capital given its capital structure (debt and equity) is the weighted average cost of capital (WACC). This is the appropriate discount rate, R.

Having presented NPV above as the only and best choice ignores that practitioners are often looking for simpler calculations. By simpler, this may be not only computationally easier, but also easier to communicate with other decision makers. Graham and Harvey (2001) surveyed corporate decision makers and found that NPV and IRR were employed roughly 75% of the time and payback 57%. The popularity of the payback method is in its simplicity (no cost of capital calculations and all cash flows ignore the time value of money or discounting). These simpler or alternative methods are discussed next in terms of their shortcomings in general finance and within a sustainability context.

Payback Period

Payback ignores the time value of money or present value of the cash flows and simply relies on when the initial investment is returned on a cash basis. Find the time, T, when the positive cash flows match the initial investments, where C_0 is negative or the initial investment outlay.

$$0 = C_0 + \sum_{t=1}^{T} C_T$$

Payback decisions rely on a stated minimum payback period. Any project or decision whose payback is beyond that cutoff would be rejected regardless of the NPV or value to shareholders.

In terms of sustainability decisions, payback-based decisions are skewed to shorter-term projects. Much criticism has been levied against US corporate behavior being too short term oriented and often focused on quarterly earnings reporting. Given one concern of sustainability is the environmental impacts on future generations, the payback method further hampers sustainable decisions. Sustainable decisions, since they take future stakeholders into consideration, would be rejected more often under the payback method. So not only is NPV a better financial method as argued by finance texts, but NPV has advantages over payback with respect to sustainability related issues and projects.

Probably one of the biggest hurdles to sustainable decision making is the short-term incentives to corporations and their managers. The NPV rule is often overlaid with other criteria such as combining payback periods with NPV for discounted payback periods. Now, discounted payback is the time that the project takes to recoup its initial investment in present value terms. Discounted payback still runs counter to the idea of sustainability with respect to future generations as well as being counter to the NPV rule.

A simple example should suffice to clarify the conflicts. The choice of a new machine comes down to the current model at $6000 producing net benefits of $2000 per year 6 years or a more sustainable model at $10,000 producing net benefits of $2500 per year for 10 years. On cost alone, the sustainable machine is higher. Assuming a discount rate of 10% for both, the net present value of the sustainable machine is $5361.42 versus $2710.52. Since this is a mutually exclusive decision, one machine or the other, the choice is to take the higher NPV or more sustainable machine. On a payback basis, the sustainable machine takes 4 years to recoup its initial cost versus only 3 years for the current machine. Even on a discounted payback basis, the sustainable machine takes over 5 years while the current machine over 3 years. A short-termed decision may be made due to employing payback or discounted payback periods of 3 years (Tables 5.1 and 5.2). The analysis also included a measure of equivalent annual cost (EAC) as a variant of NPV as well as internal rate of return (IRR).

It may be obvious that incentives for short term results from executive pay to quarterly reporting may harm the long-run performance of a firm and consequently the shareholders as well. For sustainability, even a more stringent adherence to the NPV rule may assist both shareholders and stakeholders. The case for sustainable decisions, those including more stakeholders, becomes even stronger.

Table 5.1 Less sustainable option

Year	Cash flow	PV(CF)	Cumulative DCF
0	$ (6000)		
1	$ 2000	$ 1818	$ (4182)
2	$ 2000	$ 1653	$ (2529)
3	$ 2000	$ 1503	$ (1026)
4	$ 2000	$ 1366	$ 340
5	$ 2000	$ 1242	
6	$ 2000	$ 1129	
	NPV	$ 2711	
	Payback	3 years	4 years
	EAC	-$622.36	
	IRR	24%	

Table 5.2 More sustainable option

Year	Cash flow	PV(CF)	Cumulative DCF
0	$ (10,000)		
1	$ 2500	$ 2273	$ (7727)
2	$ 2500	$ 2066	$ (5661)
3	$ 2500	$ 1878	$ (3783)
4	$ 2500	$ 1708	$ (2075)
5	$ 2500	$ 1552	$ (523)
6	$ 2500	$ 1411	$ 888
7	$ 2500	$ 1283	
8	$ 2500	$ 1166	
9	$ 2500	$ 1060	
10	$ 2500	$ 964	
	NPV	$ 5361	
	Payback	4 years	6 years
	EAC	-$872.55	
	IRR	21%	

Internal Rate of Return (IRR)

First, IRR is the rate that sets NPV equal to zero. Decisions are made based on whether the IRR is greater than or less than a required rate of return. The required rate of return may often be the appropriate cost of capital for a similar project, thus IRR will often lead to similar decisions as NPV. The two shortcomings of IRR are (1) mutually exclusive decisions and (2) non-conventional cash flows which may create more than one IRR. Concentrating on mutually exclusive projects and the relationship to sustainable options is critical to decision making. With mutually exclusive projects, it may be that a higher IRR may represent a lower NPV and therefore an inconsistent decision relative to NPV. The examples above have the less sustainable machine having a higher IRR, but lower NPV.

Mathematically, IRR is the rate, R, at which NPV = 0.

$$\text{NPV} = -C_0 + \sum_{t=1}^{T} \frac{C_t}{(1+R)^t} = 0$$

Equivalent Annual Cost: Long Versus Short Term

A variant of NPV that is important for projects of different lifetimes (years of implementation and renewal) is Equivalent annual cost (EAC). EAC bridges some of the issues arising in payback but within the framework of NPV. EAC as a variation on NPV is particularly useful in the context of mutually exclusive project analysis. While IRR is often considered the second-best alternative to NPV, one of its major shortcomings has a direct impact on sustainability decisions. EAC as another manipulation of NPV is a better alternative than IRR. Instead of solving for C_0, C_t, or R, EAC solves for the one cash flow, C, that is equivalent to the uneven set of cash flows of the project. The decision maker should choose the project with the lower EAC.

If we return to the earlier example of the two machines, on an EAC basis the sustainable machine has an equivalent annual benefit (-EAC) of $872.55 versus $622.36. The EAC analysis solves for the equivalent annual cost, C_E, such that

$$-C_0 + \sum_{t=1}^{T} \frac{C_t}{(1+R)^t} = \sum_{t=1}^{T} \frac{C_E}{(1+R)^t}$$

The projects are then compared to their relative EACs. The project with the lower EAC being accepted or higher benefit is accepted.

For the typical financial analysis a comparison of two competing processes or products comes down to analyzing the relevant costs and benefits through time to calculate the net present value (NPV). The

EAC approach does this for each investment choice with varying lives to determine the least cost alternative on an annual basis.

Sustainable Net Present Value (SNPV)

So far the discussion on financial decisions has relied on NPV employing a discount rate that is a function of expected real returns, expected inflation, and a premium for taking on risk. These all pertain to the future cash flows or consumption of a single economic agent and the trade-offs of current versus future consumption. In isolation, this works. In a society where an economic agent's consumption patterns affect not only their own utility and consumption, but also affects others' consumption patterns from today into the distant future a new or different model is necessary. Because of externalities and empathy, sustainability mandates the need for a new model. This is the one step that the SNPV model addresses by discounting effects on future generations and other stakeholders.

In corporate finance and management literature, the issue of single-agent and consumption objectives versus societal consumption objectives is highlighted through the Friedman "maximize shareholder value" objective versus stakeholder value. In sustainability, as seen in the discount rate debate, there is intragenerational and intergenerational issues of valuation. As a simplification, social distance (Becker 1968) as a broad measure of stakeholder importance to the business is adopted. While social distance employed in Anderson and Myers (2018) can be translated into cash flows that do or do not reach the intended stakeholders. Those losses in a business context may be to middlemen and in a non-profit to non-mission-related activities.

Having examined projects with traditional methods, the next step is the addition of social distance within the context of a sustainable NPV to the traditional NPV process. To avoid double counting the cash flows to the shareholders from the stakeholders, it is incumbent on the decision maker to credit any sustainable externalities, positive or negative, and their cash flows to the firm where appropriate. If sales increase from the positive marketing effects of being more sustainable, then that benefit should be recognized in the traditional NPV. Similarly, if poor press and decreased sales result from environmental damage or reputational damage, then those should also be captured in the correct modeling of cash flows in the traditional NPV analysis. The SNPV cash flows are those to other stakeholders: customers, employees, community, and future generations. In the context of a SNPV, the decision maker must choose whether or not to accept projects that have positive NPV and SNPV (think double bottom line of profit and people remembering that the triple bottom line is captured in SNPV for the environmental effects on future generations).

We define SNPV as the double summation of groups $j = 1$ to N and time $t = 1$ to T with social distances, δ_j. Shareholders are assumed to have a social distance of zero, $\delta_j = 0$. NPV is simply SNPV for shareholders only or all other stakeholders have an infinite social distance, thus making impacts or cash flows to non-shareholders zero. For SNPV, the most general form would be for each group, j, with a social distance δ_j, the SNPV function is

$$\text{SNPV} = \sum_{j=1}^{N} \frac{C_{j0}}{(1+\partial_j)} + \sum_{j=1}^{N} \sum_{t=1}^{T} \frac{C_{jt}}{(1+R+\partial_j)^t}$$

Decisions will be made based on who the stakeholders are and their social benefits. If the decision is couched in terms of different business organizations, such as a traditional public corporation versus a B-corporation, then it is possible to discover the importance of the stakeholders. Consider two projects in which the analysis has been carried out to value a traditional net present value and a sustainable NPV. The sustainable NPV (SNPV) has employed the appropriate social distance measures. A simplified example is shown below.

$$\text{NPV}_0 = -C_0 + \sum_{t=1}^{T} \frac{C_t}{(1+R)^t}$$

In the context of shareholders only, the firm would concentrate on NPV. For the inclusion of stakeholders, the SNPV form would apply (Table 5.3).

Table 5.3 NPV versus SNPV decisions

	Project A	Project B
NPV	$5000	$7000
SNPV	$3000	−$1

A stakeholder focused or sustainable NPV analysis would accept project A but reject project B. A traditional NPV decision would accept both projects A and B. Calculating the NPV and SNPV for both projects results in accepting both projects under NPV, but not under SNPV. It is obvious from the example that the addition of social distance in discounting (increasing the discount rates) lowers NPV and SNPVs. Thus, fewer projects may be accepted resulting in lower growth for the business. The lower growth from fewer projects may benefit future generations and other stakeholders more than the acceptance of marginally positive NPV projects that are not as sustainable.

The drivers for a negative SNPV versus a positive NPV for project B would indicate that the present value of the costs to stakeholders outweighs the positive present value of the benefits to shareholders. This would be an indication that the externalities were strongly negative. In terms of sustainability, those could be pollution and its negative health effects. While the majority of environmental or sustainable decisions may appear to be that NPV is greater than or equal to SNPV, due to negative externalities, this need not be the case.

In a charity, one might take on projects that benefit stakeholders "enough," where enough makes SNPV > 0. This would be indicative of a project that a typical corporation would reject, but a sustainably oriented organization would accept, SNPV > 0. In the example below, a shareholder only approach (NPV) would accept D and reject C, while an SNPV approach would accept C and reject D (Table 5.4).

The positive SNPV and positive NPV box is straightforward acceptance. The double bottom line goal of good for shareholders and stakeholders is met. The negative SNPV and NPV box decision is also a straightforward rejection, no winners. The real decision making for sustainability is the positive/negative and negative/positive boxes. But if the decision maker has correctly defined the cash flows with the appropriate social distance for each stakeholder, then the only decision is SNPV > 0. This translates to an assumption that if all economic agents are

Table 5.4 NPV versus SNPV decisions for charity		Project C	Project D
	NPV	−$2000	$7000
	SNPV	$1000	−$1

currently making optimal decisions, then all accepted investment projects have SNPV > 0.

Manipulations to the Cash Flow Model

In traditional financial decision making and especially in valuing equity, there is a going concern assumption which implies that a company lives forever, it is a going concern. If a company lasts forever, then an infinite set of cash flows must be relevant to the valuation of the company. The issue then becomes a mathematical one of valuing an infinite set of cash flows. If a company has an initial cash flow (typically assumed to be a dividend) that was just paid, C_0, and the dividend grows at a constant rate, g, forever and there is assumed a constant discount rate, R, then the value of the company's equity today is V_0. The growing perpetuity model or Gordon Growth Model in finance is written as

$$V_0 = \sum_{t=1}^{\infty} \frac{C_0(1+g)^t}{(1+R)^t} = \frac{C_1}{R-g} = \frac{C_0(1+g)}{R-g}$$

The Gordon Growth Model is most often employed in valuation, not for the value today, but the value at some point in the future, V_N. It is obvious that it would not be possible to calculate an infinite set of changing cash flows, so the financial analysis says at some point, N, the assumption is made that growth and discount rates continue at the same rates forever.

$$V_N = \sum_{t=1}^{\infty} \frac{C_N(1+g)^t}{(1+R)^t} = \frac{C_{N+1}}{R-g} = \frac{C_N(1+g)}{R-g}$$

And then substituted into the present value formula,

$$V_0 = \sum_{t=1}^{N} \frac{C_t}{(1+R)^t} + \frac{V_N}{(1+R)^N}$$

This formulation will still be applicable to SNPV analyses with the appropriate social distances such that the discount rate, R, would include both the IMRS, r, and social distance, δ. The cash flows must also represent the incremental cash flows to each group of stakeholders and the shareholders.

Business Forms

Earlier decisions about shareholders versus stakeholders have been couched in terms of the organizational structure. In this section, a deeper analysis of the organizational structure is undertaken to highlight the impact of structure on sustainability decisions. The main forms of business operations discussed in most corporate textbooks in the United States are limited liability corporations (LLCs)—both private and public, sole proprietorships, and partnerships. As a sustainability text, it is important to add non-profits and non-governmental organizations (NGOs) and B-corporations. There may be different implications of each type of organizational structure on the cost of capital, the social distances to different constituencies or stakeholders, and the overall objective of the organization in terms of profit and sustainability. The structures are also important to the size and scope of the organization.

An interesting example of the issue of size and scope arises in the structure of microfinance organizations. Some of the early microfinance organizations were structured as non-profits that provided seed money to individuals and groups to start businesses with the hope not of a profit or return to the lender, but that the growth of the businesses may fuel more growth in the community. The non-profit structure was limited to the size of the donor pool (social distance of donors to the geographical poor communities). The advantage of the initial structure of the non-profit microfinance organizations was that they did not have the tension of profit versus social or sustainable mission. Without that tension, they were able to make donations or loans and find that the default rates among groups of women entrepreneurs were lower than expected as the initial work by the Grameen Bank in Bangladesh found.

Think about how this played out in a binomial tree scenario. Lenders had been reluctant to lend thinking that the probability of default was high and therefore the expected value was low. After the NGOs showed that the probability of default was low, they also showed that the expected value was higher than traditional lenders had thought. This opened the door to larger pools of capital and commercialization of the microfinance business. The profit-oriented firms and banks that came later have targeted the more profitable sectors of the microfinance business leaving the less profitable sectors to the NGOs. The separation of the industry along organizational structure is along the lines of profitability (for profit and not for profit organizations).

A rough ranking of the size of the organizations could be sole proprietorships and non-profits then to partnerships, B-corporations, and finally to LLCs private and public. The ranking is definitely fluid across individual examples, but truer across average values. In terms of sustainability or social distance, a rough ranking might be non-profits, B-corporations, sole proprietorships to partnerships, LLCs private and public. The size and sustainability rankings would be reflective in differing costs of capital from different levels of risk and different ranges of access to capital. Thus, structure of the organization should play an important role in the objectives and goals.

Moving back toward corporations from charities and NGOs highlights the dynamic nature of the social distance framework to different types of organizations. The flexibility of organizational/business design from non-governmental organization, charities and foundations, sole proprietorship, partnership, to Limited Liability Corporation can be mapped to the different social distances in the objectives of the organizations. Charities would have lower social distances to stakeholders, B-corporations have the dual objective of financial and social that would have the lower social distances to stakeholders, and corporations having the zero social distance to shareholders and high social distances to stakeholders. Friedman's maximize shareholder wealth is equivalent to infinite social distance to non-shareholders (Table 5.5).

Table 5.5 Organizational structure with social distance

Structural form	Social distance
Charity (foundation, NGO)	Low to mission-related stakeholders
Sole proprietorship	Zero to proprietor, low to proprietor's stakeholders
Partnership	Zero to partners, potentially low to partners' stakeholders
B-Corporation	Zero to shareholders and low to stakeholders
Corporation (LLCs)	Zero to shareholders (Friedman), High to stakeholders

Sole Proprietorships

The typical corporate finance text discusses the advantages and disadvantages of different forms of business. The simplest structure is the sole proprietorship. The founder/owner of a sole proprietorship has one major disadvantage and that is personal liability. In the case of financial or legal liability, the assets of the sole proprietor are subject to seizure.

Other structural results of choosing a sole proprietorship are the life of the organization and taxes. With a sole proprietor, the life of the organization is tied to the life of the owner. It is important therefore to set up a succession plan. Similar to the life span of the organization, taxes are based on personal tax schedules given the size and profitability of the organization.

In the context of sustainability, there is another major advantage to adopting a sole proprietor structure and that is in terms of the goals of the owner and their beliefs on social distance to different stakeholders. Without the conflicts of agency issues between the owners and the managers or even among owners, there is full agreement on social distance. Think back to the Edgeworth box examples of differing utility functions as a way of explaing the conflicts with more than one economic agent.

With only the personal assets and risks of the sole proprietor, the appropriate cost of capital will on average be higher than most other forms of organizations. The capital structure of a sole proprietorship is likely to drive the cost of capital to be the highest on average. Bank loans and private debt combined may create higher costs of debt.

Partnerships

As with sole proprietorships, partnerships are treated as individuals in terms of the life of the organization, personal taxes, and personal liabilities. Initial advantages of partnership may be ease of structure during creation along with taxes. The dissolution of a partnership may be messy if not planned for in advance (think along the lines of divorce with and without a pre-nuptial agreement).

By having more than one individual and possibly many, the financial risks of partnership may be lower than that of a sole proprietorship. Emphasis on risk may be lower given more partners wealth. This may then result in a lower cost of capital.

Another potential benefit of a partnership would be if the partners are of a similar mind about sustainability and profits. Here, there may be less agreement than a sole proprietorship but more than a larger public organization or firm.

Limited Liability Corporations: Private

In the traditional finance text, most of the analysis is based on public corporations with shareholders. It is from this form that the objective of maximize shareholder value from Friedman is based. Before public firms are address, the progression of business forms here will cover private corporations. Given that one of the issues that has been raised in other organizational structures is the agreement on social distance and stakeholders, the advantage to private and to some extent closely held public firms (more than 50% ownership in one party's hands) is that stakeholder issues may be less contentious. A quick example may be a family-owned organization as an intermediate step between partnerships and public corporations as long as there is agreement within the family about which stakeholders to emphasize with a lower social distance.

Data and analysis of private firms are not as easily accessible as that of public firms. There are some conclusions that may be drawn in terms of the cost of capital relative to peers of similar industry and size. Tax affects will be similar. Cost of debt and equity may be affected by both differences in private markets versus public markets. Public markets would argue in favor of lower costs since there is a greater supply of capital.

There should be little difference in the assumed life of privately held LLC and public LLC. Practically though issues of potential takeover may be affected by public versus private markets.

Limited Liability Corporations: Public

The business or organizational form central to financial literature is the public corporation. Given the diverse set of shareholders and the pressures of public markets on profits, this form is potentially the least sustainable. If the corporation truly only cared for the shareholders in the absence of other stakeholders, then that would be the case. As has been pointed out in earlier chapters, there are advantages to firms in taking into account stakeholders such as employees and community. Management research very much favors the diverse workforce which would be consistent with UN SDGs on gender and racial equality.

Just as important are efforts by corporations to create and market products that may also support the UN SDGs for current and future generations. As governments become more involved in issues of climate change, there may also be opportunities for firms to take advantage and profit from those opportunities.

Non-profits and Non-governmental Organizations (NGOs)

Non-profit organizations such as foundations, educational institutions, and non-governmental organizations (NGOs) have two distinct characteristics to their structure that relate specifically to sustainability. First, they are typically mission driven and that mission is related to stakeholders not shareholders. Stakeholders may include specific communities, donors, employees, and recipients related to the mission. The mission and recipients may represent intragenerational and intergenerational transfers. The American Red Cross disaster relief funds are an intragenerational transfer where an organization such as the Nature Conservancy may be aimed more at intergenerational transfers. Because of the common mission, there should be less strive or disagreement about the social distance to the intragenerational and intergenerational cash flows.

There may be advantages in setting up and running a non-profit that may not be available to other business organizations.

In many cases in the United States, non-profit organizations are tax-exempt. This has implications for the required rate of return or appropriate discount rate. The cost of raising funds for a non-profit will be different than a corporation both from the source of funds and its taxation. With a different level of required returns, different sets of projects will be accepted. In the extreme, the intragenerational wealth transfer from donors to recipients need not have a positive return. The simple gift/transfer has a societal benefit to the less fortunate.

The cost of capital for a non-profit organization may be more relatable to how it fits with sustainable NPV. Cost of "equity" may be driven by the non-profit's efficiency of getting money to the appropriate stakeholders. This efficiency represents the social distances of the mission to various stakeholders. Concerns of high pay to non-profit executives may be couched in terms of social distance to the executives compared to the stakeholders for which the mission was created. Donors may examine

IRS 990 forms for this effectiveness of where the money is going and to whom.

In terms of non-profit fund-raising, the efficient use of fund-raising has an impact on donor views of the efficiency of the organization to the mission and the stakeholders with lower social distance for the donors. The simple measure of percentage of funds going toward mission serves as a first stab at measuring such efficiency. In a financial respect, the efficiency of fund-raising also has a more direct relationship to the cost of capital (the cost of raising funds is a component of the cost of capital for the non-profit). A weighted average cost of capital for a non-profit may include the percentages of funds from donors, loans from financial firms, and the funds spun off from an endowment.

The life of a non-profit is similar to that of a corporation. Theoretically, the organization could live on in perpetuity. Educational institutions' endowments are invested as if the institution will live on in perpetuity. The intriguing issue of social distance is the tension in the organization between current and future generations and ensuring that the organization is able to serve future generations without sacrificing too much to current stakeholders.

On the spectrum of sustainability, it is clear that non-profits are in general at the more sustainable end than the traditional corporation in terms of shareholder to stakeholder objectives.

B-Corporations

In between the non-profit and the corporation on the spectrum of sustainability is a relatively recent creation, the B-corporation, which is intended to be a hybrid between the two forms. A B-Corporation or benefit corporation was created to serve both profit and mission. Visit https://www.bcorporation.net/ for more information. The impact on investment decisions should be that the cost of capital must lie between the two types of organizations. This allows B-corporations to not be as restrictive on shareholder maximization and to allow some projects to move forward if there are benefits to other stakeholders.

Other Potential Advantages to Sustainable Corporations

One area of corporate finance research with respect to sustainability is the impact of more sustainably run firms on their cost of capital. In discounted cash flow models of valuation, the appropriate discount rate is key—the higher the cost of capital, then the lower the valuation. In an all-equity firm, the cost of equity may be approximated using the Capital Asset Pricing Model (CAPM). The riskiness of the firm with respect to market risk is measured by β. The riskier the firm the higher the β and the higher the cost of equity.

$$Ke = rf + \beta(\text{market risk premium})$$

If more sustainably managed firms are less risky, then they would have a lower relative cost of equity. This would translate not only into higher valuations (lower discount rate, higher valuation), but would also translate into the greater acceptance of investment projects (lower discount rate, higher NPVs, and more positive NPV projects being accepted).

Governmental Roles and International Differences

Much of the organization structure discussed above is defined in law, regulation, and tax code. Some aimed at tax consequences, some laws at governmental objectives. Organizations have the choice both to fit into the current structures to take advantages of the governmental incentives and to lobby for changes to the regulations and incentives. Societal costs and benefits may accrue from these choices. Through the last two presidential administrations, there have been a number of contradictory changes to incentives and regulations relating to sustainability. The attempt to court the Iowa corn farmers highlights the conflicting nature of governmental intervention. By approving and supporting ethanol use in fuel, the Congressional Budget Office report in April 2009 found the unintended consequences of global food/corn prices affecting lower-income household budgets as well as environmental effects. Regulatory relaxation of ethanol percentages in fuel in the fall of 2018 (Forbes, October 9, 2018) also will affect smog in the summer months. Previously, the EPA had restricted E15 fuel in the summer as it violates smog-regulations. Internationally, governments have used incentives and taxes to direct businesses and organizations for more societal goals. Subsidies for alternative energy or electric vehicles or even carbon taxes are just some of the different mechanisms employed.

Refresher on General Finance Variables

To this point, most of the variables beyond the relevant cash flows have been assumed to be given. In more in-depth modeling, it is important to understand the assumptions behind those variables. To a typical finance student or researcher, this section will serve as a gentle and simple reminder of those assumptions. To the sustainable oriented student or researcher this section provides the background to general finance necessary as a foundation underlying both finance and sustainable finance. Given that the SNPV formula relies on cash flows, C_t, and returns, R, most of what follows brings a better understanding of appropriate discount rates, R, with a bit of coverage of the growth assumptions for the cash flows, g.

General Asset Pricing Models (APM)

The most widely employed model in finance texts for valuing the cost of equity financing is the Capital Asset Pricing Model. Prior to this point, the discount rate discussed has been either a risk-free rate or an appropriate discount rate given risk. Risk-free rates are appropriate in the analysis of riskless investments that is investments with certain cash flow and timing. Typically, a riskless investment would be a US Treasury bill, note, or bond, where the differences are with respect to the length or maturity of the investment. Treasury bills are less than one year, notes between one and ten years, and bonds 10–30 years.

Given that most investments are uncertain, there is risk. Asset pricing models include the risk-free rate and some increasing function of risk or risk factors. The Capital Asset Pricing Model has only one risk factor or premium, RP, the market risk premium, MRP. An investment's sensitivity to a risk premium is typically denoted by beta. The general one-factor model representation is

$$E(R_i) = R_f + \beta_i RP_i$$

Capital Asset Pricing Model (CAPM)

The Capital Asset Pricing Model is an equilibrium, one-period model with variables of a risk-free rate, beta of asset i with the market, and the market risk premium. β may be found by regressing the excess return of the asset (R_i-R_f) on the excess return on the market (R_m-R_f). Typically, three to five years of monthly data may be used to estimate β. The market risk premium is assumed to be positive in expectations, that is consistent with investors being risk-averse. Risk aversion demands a higher expected return for taking on greater risk. Surveys and estimates have shown market risk premium in the United States to be between 3 and 8%.

$$E(R_i) = R_f + \beta_i(E(R_M) - R_f) = 2 + 0.88(5) = 6.4$$

If the beta for Starbucks is 0.88, the market risk premium is 5%, and the current risk-free rate is 2%, then an investor should demand in equilibrium to receive a return of 6.4% from an equity investment in the firm.

Taking a step back, the nominal risk-free, R_f, rate according to the Fisher Approximation is a function of the real rate of return, r, and expected inflation, i.e. nominal = real + inflation. This is a sufficient approximation for short time periods and low inflation. For risky investments, it will be necessary to employ an appropriate discount rate from an asset pricing model that includes risk or risk factors.

An all-equity firm (no debt) would use the cost of equity. The simplest and most common choice for cost of equity is the Capital Asset Pricing Model or CAPM which includes a risk-free rate (typically US Treasuries for US businesses), a market risk premium (the expected return of the market less the risk-free rate), and the business's sensitivity to market or systematic risk (beta of the firm).

$$R = E(R_i) = R_f + \beta_i(E(R_m - R_f))$$

Most asset pricing models tend to have linear risk factors are thus in the form of sensitivities, betas, and the risk factors. A model with three risk factors would be

$$R = E(R_i) = R_f + \beta_1 R_1 + \beta_2 R_2 + \beta_3 R_3$$

The next complexity to add to finding the appropriate discount rate is to account for different types of financing for the firm—equity and debt. A levered firm, a firm with debt, would employ a weighted average cost of capital (WACC) as the discount rate. WACC is calculated with the relevant weights in equity and debt and their relevant costs of equity and debt. If the organization is taxable, then the cost of debt is typically the after-tax cost of debt since debt is a considered as a cost to the organization.

$$R = \text{WACC} = w_e R_e + (1 - \tau) w_d R_d$$

where w_e and w_d are the respective weights of equity and debt, τ is the marginal tax rate, and R_e and R_d are the respective costs of equity and debt. NPV accounts for the timing of the cash flows and their present value via the discounted cash flows. Risk of the investment is captured in the cost of capital, the discount rate, R.

Within the discount rate assumptions, there may be assumptions about inflation's effect on risk-free rates which would have a direct effect on the expected nominal return. For the most problems in low inflation environments, currently most of the developed world, the Fisher Approximation is sufficient. The Fisher Approximation is nominal interest rates are approximately equal to real rates of return plus expected inflation. Continuing with simplifying assumptions, a constant real rate of return assumption removes the issue of trying to estimate future real rates of return and has the added benefit of assisting in removing most arbitrage issues across interest rates. If real rates of return are constant globally, then currency futures are just a function of differences in expected inflation across countries and currencies (Table 5.6).

Building on the assumptions above provides the following simplified equations for arriving at the appropriate discount rate.

Fisher Approximation Nominal $=$ real $+$ inflation $= r + i = R_f$
CAPM with fixed MRP $E(R_i) = R_f + \beta_i(\text{MRP}) = R$
Growth in cash flows is the retention ratio times the return on equity, $g = b(\text{ROE})$

Table 5.6 Variables fixed and random

Fixed	Random
MRP $= E(R_i) - R_f$	Cash flows, C_t
Retention ratio $= b$	Inflation, i
Real rate of return, r	Return on equity, ROE
Beta, β	

Growing perpetuity,

$$V_0 = \sum_{t=1}^{\infty} \frac{C_0(1+g)^t}{(1+R)^t} = \frac{C_1}{R-g} = \frac{C_0(1+g)}{R-g} = \frac{C_0(1+\text{bROE})}{R-\text{bROE}}.$$

References

Anderson, A. M., & Myers, D. H. (2018, Winter/Spring). Sustainability: Discounting the Future, Social Distance, and Efficiency Effects. *Moral Cents*.

Becker, G. (1968, March/April). Crime and Punishment: An Economic Approach. *Journal of Political Economy, 76*(2), 169–217.

Graham, J. R., & Harvey, C. R. (2001). The Theory and Practice of Corporate Finance: Evidence from the Field. *Journal of Financial Economics, 60*.

CHAPTER 6

Investment Implementation

Abstract This chapter provides a review of the investment opportunities and strategies for sustainable-minded investors. A review of the historical sustainable movement from ethical screens to socially responsible investment to mission related and impact invest investing highlights changes in the costs of monitoring and investor interest that have created a greater funds flowing to sustainable investing. Strategies from positive and negative screening to shareholder activism as well as best in class are discussed.

Keywords Investment · Sustainability · Social distance · ESG · CSR

This chapter covers investment opportunities and strategies for sustainable-minded investors. Additionally, the role of modern portfolio theory and asset pricing gives context to where sustainable investing fits in the academic literature and research. A perspective of the history of investment theory, information, and technology provides an understanding of the changes and growth in the wide range of sustainable investment approaches. In particular, a review of the historical investing sustainable movement from ethical screens to socially responsible investment (SRI) to mission-related (MRI) and impact investing highlights the changes in the costs of monitoring and measurement along with investor interest that have helped create a greater involvement and more funds in sustainable investing. The efficiencies in measurement of sustainable criteria through lower costs and the increase in the acceptance and trust of employing those sustainable measures for decision making have aided in the expansion of sustainable investing. The past forty years from 1980–2020 have seen the confluence of computerization of the investment business and the growth in the number of firms measuring environmental, social, and governance (ESG) issues.

THEORY TO EVOLUTION IN PRACTICE

By keeping in mind the underlying economic truths, it is easier to understand the changes in the investment opportunities. If an investor examined the investment opportunity set from the 1950s to today, they would observe an explosion of opportunities in type and scale. In the 1950s, the majority of stocks were held by individuals and not institutions. Pension funds and mutual funds were invested in balanced accounts (both stocks and bonds). In the 1970s with the advent of the Employee Retirement Insurance Security Act 1974 (ERISA) and its mandate for diversification and fiduciary responsibility, the growth in separate equity and bond accounts took off with the help of academic research. Markowitz portfolio theory and Sharpe's CAPM in the 1950s and 1960s drove much of this change in diversification. Diversification was later codified in pension fund investing in the 1970s with the ERISA.

The diversification drive was met with the expansion of equity investment opportunities. As with perfect markets, product differentiation came in the form of value and growth stock portfolios as well in market capitalization size portfolios. This was followed by relaxation in international capital investment taxes in the US in the 1970s and regulation opening foreign markets to large investors.

The nineteen eighties (1980–89) ushered in growth in defined contribution plans (DC plans). The cost of tracking investment amounts for the vast number of individuals in defined contribution plans was driven lower thanks to computer memory and processing costs coming down. Further momentum towards DC plans was created by FASB 87 requiring defined benefit (DB) plan underfunding be shown on corporate balance sheets as a liability (increasing the marginal cost of DB plans to corporations). A parallel occurrence in lowering the cost of ethical investing information began with the South Africa divestment movement in the 1980s. As large endowments and public pension plans demanded the information, suppliers of the information, IRRC, came into being.

The nineteen nineties (1990–99) investment advances both sustainable and not were greatly assisted by the continued advancement in computer processing and storage costs falling along with the expansion of the Internet bringing information costs down. On the sustainability side of the information efficiencies, KLD in 1991 began publishing its socially responsible investing data, thus assisting in the decrease in information costs for all SRI investors. The introduction of Exchange Traded Funds (ETFs) (SPRS in 1995 as an S&P 500 index fund available to trade intraday) further lowered investment costs. The digitalization of equity trades from an eighth of a dollar to pennies assisted the lowering of costs across the equity world.

The growth in sustainable investing has been greatly aided by the growth, efficiency, and transparency of the data necessary to make better sustainable decisions. From the IRRC's data (1980s) on South African divestment combined with the Sullivan Principles as the criteria of measurement to the KLD data (1990s) and indices to sustainable reporting (2000s), there has been tremendous growth in data, indices, consultants, and governments using sustainability criteria. Companies in many countries in Europe along with Australia and New Zealand must provide sustainability reports. Even in the United States where there is no requirement, more firms are now producing sustainability reports for their investors. This growth in information has followed, mirrored, or just correlated with the growth in funds tied to sustainable criteria. The United Nations Principles of Responsible Investing (UN PRI) has signatories representing trillions of dollars of assets under management https://www.unpri.org/about/the-six-principles. "They have more than 1750 signatories, from over 50 countries, representing approximately US$70 trillion" (Table 6.1).

The UN PRI in its definition of responsible investing states that while it has similarities with SRI, ESG, impact, sustainable, green, and other types of investing that responsible investing is broader. "…whereas responsible investment is a holistic approach that aims to include any information that could be material to investment performance." (https://www.unpri.org/about/what-is-responsible-investment) The UN PRI's broader definition does allow for greater acceptance than the narrower definitions of some of the other aspects of sustainable investing.

Table 6.1 UN PRI 6 principles

Principle 1: Incorporate ESG issues into investment analysis and decision-making processes
Principle 2: Be active owners and incorporate ESG issues into ownership policies and practices
Principle 3: Seek appropriate disclosure on ESG issues by the entities in which we invest
Principle 4: Promote acceptance and implementation within the investment industry
Principle 5: Work together to enhance our effectiveness in implementing
Principle 6: Report on our activities and progress

The growth in sustainable investing has been mirrored in the academic literature over this period with studies of Corporate Social Responsibility (CSR) emanating from the management literature as well as accounting and finance studies on corporate and investment performance. Friede et al. (2015) provide a meta-analysis of over 2000 research reports on sustainable investing. Across the 2000 reports, there have been mixed results of the performance of sustainably managed funds. One conclusion of the mixed results is that given the uncertainty of investing, there is no conclusive advantage or disadvantage to following sustainable criteria. This conclusion is supported by Anderson and Myers (2007) given the theoretical ex ante notion that constrained portfolios should underperform unconstrained portfolios. Given that this underperformance is not conclusive, investment funds may follow sustainable criteria without necessarily impacting performance on a risk and return basis.

History of Sustainable Screening

It seems appropriate to begin this section of the historical background, "in the beginning" there were religious tenets and those tenets informed individuals of what was acceptable and what was not—The Ten Commandments, The Golden Rule, and Sharia Law, for example. When applied to investments, the simple cost-effective approach was thou shalt not invest in evil. In economic terms, these rules were low cost and straightforward. In portfolio choice, the translation initially was to negative screening. One removes, for example, alcohol and tobacco stocks from a portfolio. Some of the earliest examples of sustainable investing go back to religious organizations and the moral or ethical standards being applied to their investment portfolios for individual members or in modern investing in the pension plans for their employees. A Mormon fund may choose to exclude tobacco, caffeine, and other vice stocks. A Catholic fund may exclude pharmaceuticals related to birth control. This methodology is referred to as negative screening. The economic implication of negative screening is that there is an infinite cost (sin) to investing in such firms and therefore no trade-off of demanding higher monetary return to offset the "sin." In terms of theoretical SNPV, the social distance is infinite to any cash flows (returns) from sin related companies. Practically, the social distance to these cash flows does not need to be infinite. As long as the social distance is great enough to exceed the possibility of high returns justifying an investment in sin stocks.

Negative screens taken in isolation meet the needs of the righteous to maintain their moral objectives. As individual investors, those decisions are based on the individual beliefs or values. Since the 1950s, the growth of pension funds (public and private), endowments, and foundations, and the governmental regulations that grew with them and that grew out of the Great Depression (Investment Company Act of 1940) have brought the issue of risk-return trade-offs to the forefront. Department of Labor guidance of non-investment related decisions having no return implications is an example of governmental oversight on investment choices.

Now, the issue of financial economics and its related investment theories plays a role to meet governmental and legal fiduciary requirements. Returning to negative screens, two implications immediately jump out. One is that in terms of Markowitz mean-variance portfolio theory, any restricted or constrained portfolio ex ante must be less than or equal to the unconstrained portfolio in terms of risk and return. This was the aim of Anderson and Myers (2007) to test screened or constrained portfolios versus unconstrained using both positive and negative screens. An investor will not/would not put a tobacco stock in their tobacco-free portfolio at any level of expected return. Even if an infinite return were guaranteed with no risk, such an investor would not include it in their portfolio. They have implicitly assumed an infinite cost to their utility function. In the extreme, this may seem absurd, but the basis for the negative screening came originally from moral and ethical tenets. An infinite cost or damnation could be the penalty for the sin of violating those tenets. On a practical basis, this is difficult to imagine but the infinite cost is one direct interpretation of such exclusionary behavior. The advantage of negative screening was the low-cost implications for its implementation. As the cost of measurement has come down, the ability to weigh the trade-offs of the "sin" and the return is easier to make. Negative screening is being joined by more nuanced investment and sustainability techniques to investing such as lower weights to poor sustainable investments.

There was a progression from negative screens to positive screens in sustainable investing. Invest in what is good or what is better. Similar to negative screens the positive screening process was also initially low cost. This progression also coincided with the advent of sustainable investment monitoring companies. An early example is the South African divestment movement as a response to apartheid among endowments, foundations, and public pension funds. The negative screen approach was low cost and straightforward; do not invest in companies that invest or have operations in South Africa. Next came the Sullivan Principles which were more along the line of a positive screen; add back firms that are doing good in South Africa by adopting the Sullivan Principles. One such provider of screening information during the 1980s was the IRRC. The IRRC did the research, published, and sold the list of Sullivan Principle compliant firms to the investment community to meet the demand of investors wanting South Africa free or Sullivan Principle compliant portfolios. The low-cost negative screens were turned into higher cost positive screens.

The next step in investment terminology was socially responsible investing (SRI) which broadened from the more negative ethical and sin exclusion to include more issues through both negative and positive screening. The positive screening represented a push toward social good. Issues such as diversity (race and gender) and environmental allowed investors to seek companies that were involved in positive social change. As the Sullivan Principles were to divestment, so socially responsible positive screens were the next step from the initial negative screens of ethical investing.

The ability to carry out this next step came down to continued decreases in the cost of information and increased transparency. Much of this progress was brought about by firms that were researching and publishing the information which brought about ranking systems for corporate behavior and lead to the creation of socially responsible investment indices and products. KLD created one of the first SRI indices in 1990.

One outcome of the research and ranking of corporate behavior was the ability to manage the SRI indices on a best in class methodology

Table 6.2 Categories of qualitative screens from Anderson and Myers (2007)

Qualitative screens	Exclusionary screens
Community	Alcohol
Corp. governance	Firearms
Diversity	Gambling
Employee relations	Military
Environment	Nuclear power
Human rights	Tobacco
Product	

where higher ranked firms within an industry or sector received greater weight in the index than their lesser ranked industry competitors. Instead of strict positive or negative screens on sustainable issues for compliance, the criteria allow for ranking and scores for firms. A best in class system based on the ratings then affects the relative weights within an index. Different index providers and their numerous rating systems have led to some confusion and frustration in the investment industry about what is or is not a sustainable investment. A recent MIT working paper by Berg et al. (2020) attempts to explain some of the confusion and dispersion of firm ratings. The different ratings can be thought of as having different social distances which affect the ratings and the weightings of firms in a sustainable index (Table 6.2).

Returning to the emphasis of the axiom of "if you measure it you can manage it," this twenty-first century has seen the adoption by many in investment community including ESG factors in their investment process. As the market has grown for sustainable investing, so has the competition for investors and funds. Since early investment opportunities were negative screens and thus constrained portfolios, one rallying cry is that there can and is positive returning strategies that are available and that may have a positive financial and social return. Social venture funds, community banking and investments, and even sustainable private equity funds are now part of an impact investment movement of positive returns with positive sustainable impact (Table 6.3).

Investor Types

Having laid out the role of utility functions with return, risk, and social distance in earlier chapters, the differences among investor types are

Table 6.3 Investment approaches by implied social distances and information costs

Investment approach	Implied social distances	Information costs
Ethical screens	infinite cost	low
SRI screens	infinite cost	higher
Sustainable/Best in class	finite cost/changing weights	moderate information costs
ESG factors	approaching zero	low
Shareholder activism	high	low

reflected in their approach to sustainable investing. This section walks through the different investors and their different approaches. Along the spectrum of social distance are individuals with the lowest social distances, followed by foundations and endowments, and then to public pensions and finally to corporate pensions with high to infinite social distance to stakeholders.

Individuals

For individual investors who want to combine financial returns with sustainable returns, there are now a number of avenues available to them. The mutual fund industry has expanded their options for taxable accounts with ESG/SRI mutual funds and exchange trade funds.

Additionally, individuals through their employers' 403b options and 401K options have been given access to mutual funds and ETFs within their plans that are ESG/SRI.

Indirect investments such as mutual funds and ETFs already provide individual investors with lower-cost entry points into diversified investing. The denomination intermediation of allowing an investor to be exposed to hundreds or thousands of investments in one investment vehicle is the main advantage of the mutual fund industry. Combining sustainably managed funds to their suite of options has further lowered the barriers of entry to sustainable investing for the individual.

Pensions

Public

Public pension plans with investment boards consisting of public union officials, politicians, and others with political agenda have set up investment policies that support local businesses and diversity requirements of gender and race. This provides those boards another approach to support those investments that reflect their lower social distance to their constituents.

The behemoth size of California Public Employee Retirement System (CalPERS), over $400 billion in early 2020, allows them an impressive set of options relative to most other funds. CalPERS has been a leader for many years in shareholder activism. As a large investor, the option to influence shareholder proposals is available to them. CalPERS even has a Sustainable Investment Research Initiative database of over 1800 academic articles.

https://www.calpers.ca.gov/page/investments/governance/sustainable-investing/siri-library

Private

In contrast to the more political nature of public pension funds are the corporate pension funds. Corporate pensions, defined benefit plans, are a liability of the corporation (FASB 87). Optimizing the performance of the pension fund may both reduce the unfunded liability and decrease the cash flow needs from the corporation to the pension fund. Both of these outcomes would serve the Friedman mantra of maximize shareholder wealth.

With the fiduciary responsibility of the defined benefit plans covering all the beneficiaries and juxtaposed with the corporation's goals, the analysis of conflicting social distances of the beneficiaries as discussed in Chapter 2 on utility functions (see Edgeworth Box discussion) suggests that avoiding competing social distances and differing beliefs on sustainability results in corporate pension plans being much less involved in sustainable investments.

Endowments

Educational endowments may be one of the few investor groups with the longest time horizon and therefore the greatest reason to be sustainable.

In serving their institution for the long term, the social distance of the endowment to future generations must be lower than other investors to future generations in that their mission is to provide for the continuing education of this current and future generations. If there are no future generations due to environmental degradation, then there is no need for the continued sustainability (financial or otherwise) of the institution.

The stakeholders for the endowment include their alumni, current and future students, employees, and their communities. Since the "profit motive" of the endowment is to be able to support the institution, the investment horizon may be infinite. To assist in the continued existence and often to maintain tax-exempt status, there are governmental regulations and controls on the annual spending limits to ensure that the endowment continues. With a lower return threshold and a longer-term horizon, endowments are perfectly positioned to lead in the areas of sustainable investments.

Foundations

The advantage that foundations have over pension plans and especially corporate plans is that of purpose or mission. The agreement of mission may provide a more unified approach to sustainability. So much so that the classification of mission-related investment (MRI) is a perfect alignment of foundation purpose with social distance agreement.

Size and mission play a large part in the investment decisions of foundations. The more focused a foundation's mission on a particular problem (health, education, ...) the more appropriate it may be for the foundation's investment choices to be consistent with their spending. Obviously, the alignment of investment and consumption (spending) are linked within the utility function frameworks.

An interesting parallel is Kinder (2005) defines social investors into three categories—value-based, value-seeking, and value-enhancing. It is easy to place the value-seeking investors as interested in high levels of return, value-based investors as having close social distances, and value-enhancing investors as being value-seeking with very far social distance (Fig. 6.1).

- Interpreting the investor types by their location on the graph and starting in the lower left corner of lower returns and lower or close

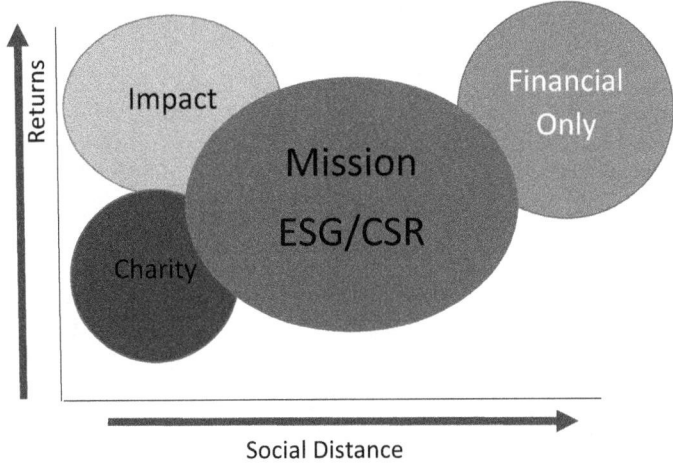

Fig. 6.1 Returns and social distance to non-shareholders (stakeholders)

social distance is Charity. Charities may have close social distance to their mission and require a moderate level of returns.
- Impact investing also requires close social distance with charities, but greater expected returns.
- In the upper right corner are investors with only financial objectives which assumes great social distance and high expected returns.
- Mission-related investing requires a combination of impact and charity with closer social distance and moderate to higher expected returns.

Another approach to examining the relationship of investor type to social distance and return is presented below (Table 6.4).

Governmental Role

Department of Labor Rulings

The Department of Labor oversees pension plans and the Employee Benefits Security Administration. Through that oversight the department issues guidance. Given different administrations that guidance has altered some of its views on ESG issues. While the Obama administration gave more leeway to pension funds, the Trump administration has pulled back

Table 6.4 Investor type with social distance and returns

	Today	Future	Distant future
Investor			
Individual	Charity	Charity	Charity
	$\delta > 0$, $Rt = 1$	$\delta > 0$, $Rt > 1$	$\delta > 0$
	Personal consumption time zero goals	Retirement, Housing, Education, Children, Grandchildren	Descendants
	$\delta = 0$	$\delta = 0$	$\delta = 0$
Pension	Current beneficiaries	Current Employees	Future employees
	$\delta = \infty$	$\delta > 0$	
Endowment	Students/Institution	Students/Institution	Students/Institution
	$\delta > 0$	$\delta > 0$	$\delta > 0$
Foundation	Mission	Mission	Mission
	$\delta > 0$	$\delta > 0$	$\delta > 0$

δ is the social distance parameter and R is the time discount factor

somewhat. In 2018, Field Assistance Bulletin No. 2018–2001 clarifies the administration's view compared to Interpretive Bulletins 2016–2001 and 2015–2001. One quick summary of the administration's view is "The Department has a similarly longstanding position that ERISA fiduciaries may not sacrifice investment returns or assume greater investment risks as a means of promoting collateral social policy goals." In the context of sustainability and finance, the approach is that investments may not sacrifice return or increase risk to promote sustainability. As long as no sacrifices are made in those directions, then sustainability, ESG, may be used as a tiebreaker by fiduciaries. With the academic research record mixed on sustainable investing, it appears there is no significant cost in terms of risk and return to sustainable investing. In June 2020, the Department of Labor has proposed more restrictions on ESG investing by private pension plans.

Choice of Sustainable Approach

As the last section, the arguments put forth come full circle. Investors have choices about how they approach markets and sustainability. Those choices, through revealed preferences, provide insight into the utility functions, belief systems, information efficiency, market forecasts, and governmental regulation of each investor. An investors' approach is dependent on their consumption goals (utility function), the constraints they face (governmental regulation or organizational structure), and their return assumptions (market forecasts) as well as their social distance goals and assumptions.

From history, the approaches to sustainable investing in all its forms have been from screening (positive and negative), to relative weighting (overweight the good and underweight the bad), to shareholder activism. As discussion earlier, much of this evolution has been driven by decreasing costs of information as popularity has grown. On the data information side, the increase in the number and scope of databases providing information about sustainable issues has been significant. Even finance.yahoo.com has ESG scores available for mutual funds. The data are provided by Morningstar since 2017, but the underlying scores are from Sustainalytics. Mutual fund scores are based on Environment, Social, and Governance measures less a measure for Controversy. The measures are combined based on asset value weights of the scores of the firms in the portfolio. For individual firm information on finance.yahoo.com, sustainability scores are also based on Environmental, Social, Governance, and Controversy. The individual firm data have been available from Sustainalytics since 2014.

For large institutions, California Public Employee Retirement System has been a leading advocate of shareholder activism. The size of large public retirement systems, endowments, and foundations gives them the leverage to push sustainability issues on to proxy votes changing corporate behavior. While it takes over 50% to pass a resolution, often corporations will respond to pressure from shareholders with less than 50%. Obviously, share prices are subject to supply and demand issues and the resulting pressures will affect a firm's cost of capital. With the increase of firms specializing in proxy voting for institutions, the marginal cost of the shareholder advocacy approach has also seen a decrease through time thanks to computers (speed and memory) and competition. The analysis here may be couched in terms of increasing efficiency of shareholder efforts or a decreasing cost/loss of such efforts.

Constrained and Unconstrained Portfolios

In terms of the simplest and earliest form of sustainable via screening, the analogous optimization issue is to constrain or limit the portfolio investment choices. In Markowitz mean-variance parlance of risk and return, the constrained (or screened) efficient frontier should lie tangent or inside the unconstrained efficient frontier. Theoretically, the ex ante choice has the market's expectations of risk and return built into them. If the market is correctly pricing sustainability factors then this constrained portfolio underperformance should hold ex post. Much of the research of sustainability investing centers on this ex ante-ex post issue. Findings have been mixed. Much of this has to do with the time periods chosen and with the random nature of returns. As Anderson and Myers (2007) find, there is "no cost to being good." The mixed results, sometimes positive and sometimes negative and often indeterminant, imply that as with most competing investment strategies, it is difficult to find consistent winners.

The other approach to the issue of constrained versus unconstrained portfolio choice is to realize that the sustainably driven investor's utility function is not simply mean and variance. Once the correction of utility functions is made the question then becomes is the mean-variance portfolio inferior to the correct sustainable portfolio for the investor. The answer ex ante is obviously yes; future and continued research should explore further if this holds true ex post. In terms of consulting to investors, it now becomes as great an effort on the discovery of an investors' sustainability (social distance) goals as it is an effort to find the best investment risk and returns.

Climate Change Investing and Risk Management

With the recent emphasis in the news of climate change or impending climate disasters, investment firms are following the news with a trend to find investment opportunities. The investment opportunities look for how climate change will affect businesses and thus affect investment returns. A quick example is the travel industry. Which areas of the world and what types of travel will benefit or suffer from climate change? Is snow skiing viable long term? Are certain beach resorts and islands susceptible to rising ocean levels?

A more in-depth analysis of climate change and investing centers on risk analysis of businesses. If there are increasing threats and financial effects from worsening storms, then there will be increasing financial risks to the property and casualty insurance business. Agricultural businesses will also be affected through changes to rainfall and temperature swings. Analysis of crop successes and failures and the resulting price affects will provide investment opportunities and risks. NPR ran a story on October

12, 2016 on the viability of coffee due to the effects of climate change in Brazil (https://www.npr.org/sections/thesalt/2016/10/12/497578413/coffee-and-climate-change-in-brazil-a-disaster-is-brewing) based on a report by The Coffee Institute, "A Brewing Storm: The climate change risks to coffee."

On July 24, 2019, the New York Times reported that Moody's had purchased a large stake in a climate data firm in order to better access the risks. Risks are not the only investment approach to climate change. There will also be investment opportunities in finding solutions and new approaches to climate change by providing increasing sustainability. As governments regulate solutions or fund research and investment in sustainability, there will be growing investment opportunities. The most visible investments have been in alternative energy sources, such as wind and solar. The growth in battery technology has supported the growth and viability of electric vehicles. That growth has been supported by local and national government subsidies and rebates for electric vehicles. In addition, many European governments have passed legislation mandating the end of internal combustion engines by 2040 or 2050, all in an effort to slow climate change. Norway has been the most aggressive in this area by moving toward ending new car sales of internal combustion engines by 2025.

Portfolio Creation

Given the variety of investors and definitions of types of sustainable investing, the creation of a sustainable portfolio may vary for each type of investor based on their definition of sustainability or social distances. This variety has caused some economies of scale issues and dampened the breadth of acceptance of sustainable investing. Think back to Chapter 2 and Edgeworth box examples for societal agreement and you can imagine trying to market a sustainable mutual fund product. There will be those investors who do not accept non-profit-maximizing criteria and then there will be investors who do not agree on the type of sustainable investing they want to do. Think about ethical investing based on religious criteria and trying to get agreement among different religious portfolio managers about what is acceptable to them (Table 6.5).

As an example, Table 6.2 has the large-cap market index fund from Vanguard and the two largest SRI funds according to USSIF—TIAA-CREF's Social Choice and Parnassus's Core Equity fund as of June 30,

Table 6.5 S&P 500 top 10 versus 2 largest SRI mutual funds

Vanguard S&P 500		TIAA-CREF social choice		Parnassus core equity		MSCI KLD 400 Social ETF	
Name	% Assets	Name	% Assets	Name	% Assets	Name	% Assets
Microsoft Corp	4.18	Microsoft Corp	3.85	Microsoft Corp	6.03	Microsoft	7.43
Apple Inc	3.52	Apple Inc	3.20	Disney	5.09	Facebook	3.50
Amazon.com	3.19	Amazon.com	2.01	Linde plc	3.93	Alphabet GOOG	2.59
Facebook	1.89	Facebook	1.76	Danaher Corp	3.55	Alphabet GOOGL	2.47
Berkshire Hathaway	1.63	Alphabet GOOG	1.45	Mastercard	3.53	VISA	2.31
Johnson & Johnson	1.51	Alphabet GOOGL	1.43	American Express	3.51	Proctor & Gamble	2.09
JPMorgan Chase	1.48	Proctor & Gamble	1.42	Clorox Co	3.21	Disney	1.91
Alphabet GOOG	1.35	Cisco Systems	1.29	Cadence Design Systems	3.18	Mastercard	1.84
Alphabet GOOGL	1.32	Verizon	1.26	Costco Wholesale	3.11	Cisco Systems	1.83
Exxon Mobil Corp	1.32	Merck & Co	1.24	Cerner Corp	3.08	Verizon	1.80
Total Top 10	21.39		18.91		38.22		27.77

(*Source* finance.yahoo.com holdings)

2019. The TIAA-CREF fund is very similar to the Vanguard fund while the Parnassus fund has only the largest holding, Microsoft, in its top ten. TIAA-CREF has more overlap with the MSCI KLD 400 Social ETF. MSCI KLD 400 index was started in 1991 as the first major socially responsible index to compete with the S&P 500 index.

OTHER INVESTMENT VEHICLES

While most of the preceding discussion has been related to equity investments. Sustainable investing now cuts across traditional and alternative investment classes. In fixed income investing and especially in Germany, there is the advent of Green bonds. Green bonds are somewhat analogous to B-Corporations in that they have dual objectives—return and environmental goals. If the underlying corporation that has issued the bonds fails to meet certain green criteria, the coupon rate on the bonds may rise or if they meet the targets the coupon rate may fall. Green bonds provide an economic incentive for the corporation to meet the sustainability targets.

Within the alternative investment universe (private equity, venture capital, and real estate), there are sustainable investment options. Many of these may be categorized as impact investing, that is the investments often also have a dual purpose to meet return targets and sustainability targets. Community investing may target low-income areas for improvements. Impact investing through venture capital or private equity may target green technology firms or startups to assist with technological innovations.

REFERENCES

Anderson, A. M., & Myers, D. H. (2007). The Cost of Being Good. *Review of Business*, (Autumn).

Berg, F., Kölbel, J., & Rigobon, R. (2020). Aggregate Confusion: The Divergence of ESG Ratings, MIT Sloan School Working Paper 5822–19.

Friede, G., Busch, T., & Bassen, A. (2015). ESG and Financial Performance: Aggregated Evidence from More than 2000 Empirical Studies. *Journal of Sustainable Finance & Investment*, 5(4), 210–233. https://doi.org/10.1080/20430795.2015.1118917.

Kinder, P. (2005, September 1). Socially Responsible Investing: An Evolving Concept in a Changing World. *KLD Research Paper*, p. 3.

Markowitz, H. (1952, March). Portfolio Selection. *Journal of Finance*, 7(1).

Epilogue

Researchers, students, organizational leaders, and others who have read this book have been given a decision-making structure using economic techniques and models to approach the inherent sustainability choices that must be made. Sustainability reflects the choices of the effects on those in current and future generations. In a business context, those affected are the shareholders, employees, customers, suppliers, and the community both present and future. The model provided in this text gives a structured understanding of the externalities' impacts both positive and negative and the relative importance to the decision maker through discounted cash flows with both a social distance measure and a time value of money discount rates. The size and effects of those decisions may not rise to the level of importance to change a decision, but without the analysis and reflection that have been presented the decision maker will not know if a more sustainable decision has been missed. Researchers can provide a broader context by examining past decisions to understand when and how decisions reflect the social distance discounting implied. This will provide direction for future decisions.

Adoption of a sustainable net present value model by businesses leaders based on the time value of money and social distance for discounting should create more sustainable decisions. Sustainable decisions reflect the understanding of the trade-offs among current stakeholders (shareholders, suppliers, customers, employees, and communities) and among future stakeholders. Some of those decisions will be made within societal norms driven sometimes by governmental regulations, but some of those decisions will be made based on the empathy and ethical standards of the individual business. It is a greater understanding of sustainability that has been reached for here.

Index

A
Affordable and clean energy, 56
asset pricing, 79

B
B-Corporations, 42, 62
Becker, Gary, vii, 6, 11, 69
binomial pricing, 32

C
Capital Asset Pricing Model (CAPM), 29, 79–82
Clean water and sanitation, 56
Climate action, 56
consumption-based utility, 25
Consumption-based utility models, 19
Corporate Social Responsibility (CSR), 82

D
Decent work and economic growth, 56

defined benefit (DB), 80, 87
defined contribution plans, 80
demographics, 41
discount rate, 28

E
Edgeworth box, 35
efficient charity, 7
efficient altruism, 29
Employee Retirement Income Security Act 1974 (ERISA), 80
endowments, 81, 83, 84, 86–88, 91
environmental economics, 11
Environmental Kuznets Curve, 47
Environment Social Governance (ESG), 7
Equivalent Annual Cost (EAC), 68
Exchange Traded Funds (ETFs), 94

F
Fair Trade, 50
Fair Trade Certification, 50
Fisher Approximation, 29

Friedman, Milton, 35

G
Gender equality, 56
Good health and well-being, 56
Gordon Growth Model, 38, 39
Great Financial Crisis, 5
growth, 41

I
impact investing, 79, 89, 95
Industry, innovation and infrastructure, 56
Internal Rate of Return (IRR), 61, 63, 65, 67, 68
Iroquois, 31
Iroquois Nation, 8

L
LEED, 50
leverage, 41
Life below water, 56
Life on land, 56

M
Markowitz, Harry, 80, 83, 91
mission related, 79, 89
mission-related investing, 88
modern portfolio theory, 79
mutual funds, 80, 86, 91

N
Net Present Value (NPV), 61, 63–71, 77, 79, 82
No poverty, 56

P
Partnerships for the goals, 56

Peace, justice and strong institutions, 56
Pension, 87
  corporate, 87
  public, 87
Pension funds, 80
Pension plans, 81, 83, 88, 89
  defined benefit (DB), 80, 87
  defined contribution, 80
Piketty, Thomas, 28
Plato's Allegory of the Cave, 14

Q
Quality education, 56

R
Reduced inequalities, 56
Responsible consumption and production, 56

S
Screening
  negative, 82–84
  positive, 84
social distance, 6, 11
socially responsible investing (SRI), 84
socially responsible investment, 79, 84
societal utility function, 10
South Africa divestment, 80
  Sullivan Principles, 81, 84
Sustainable cities and communities, 56
Sustainable Net Present Value (SNPV), 11, 61, 62, 64, 69–72, 80

T
Time (T) in Years, 13
triple bottom lines
  People, Planet, and Profit, 20

U

United Nations Principles of Responsible Investing (UN PRI), 81, 82

United Nations Sustainable Development Goals (UN SDGs), 3, 4, 6–10, 12, 13, 16, 47, 56

W

weighted average cost of capital (WACC), 45

Z

Zero hunger, 56

CPSIA information can be obtained
at www.ICGtesting.com
Printed in the USA
LVHW080717140722
723495LV00004B/191